Next-Gen Manufacturing 2021

Topical Conference at the 2021 AIChE Annual Meeting

Boston, Massachusetts, USA and Online
7 - 11 November and 15 - 19 November 2021

ISBN: 978-1-7138-5693-1

Printed from e-media with permission by:

Curran Associates, Inc.
57 Morehouse Lane
Red Hook, NY 12571

Some format issues inherent in the e-media version may also appear in this print version.

Printed with permission by Curran Associates, Inc. (2022)

For permission requests, please contact AIChE
at the address below.

AIChE
120 Wall Street, FL 23
New York, NY 10005-4020

Phone: (800) 242-4363
Fax: (203) 775-5177

www.aiche.org

Additional copies of this publication are available from:

Curran Associates, Inc.
57 Morehouse Lane
Red Hook, NY 12571 USA
Phone: 845-758-0400
Fax: 845-758-2633
Email: curran@proceedings.com
Web: www.proceedings.com

TABLE OF CONTENTS

Influence of Interfaces in Electrical Properties of 3D Printed Structures 1
Fraser Daniel, Andy Gleadall, Adarsh Radadia

Modeling Heat Transfer in Material Extrusion Additive Manufacturing: Balancing Model Resolution with Computational Demands 3
Michael Bortner, James T. Owens, Arit Das

Assessing the Fidelity of Additively Manufactured Objects 4
Hajar Taheri Afarani, Edward Garboczi, Newell H. Moser, Ebrahim Nasr-Esfahani, Joseph J. Biernacki

Design and Processing of Open Lattice Structures for Tunable Fluid Phenomena 5
Ian Woodward, Lucas Attia, Premal Patel, Catherine Fromen

Methods to Evaluate Residual Stress in FDM Printed Parts 6
Daniyal Shoukat, Connor Forte, Macrae Montgomery, Haley Hilborn, Jerry Qi, Nese Orbey

Accelerating Time to Competency in an Industry 5.0 World 7
Ian Willetts, Brent Kedzierski

How to Get a Head Start on Your Digitalization Journey 14
Jonas Norinder

Development of an Advanced Control Infrastructure for Subcritical Power Plant Online Power Demand Tracking 15
Daniel Kestering, Selorme Agbleze, David Tucker, Lawrence J. Shadle, Heleno Bispo, Fernando V. Lima

CFD Based Study to Predict Rollover in an Industrial LNG Storage Tank 16
Suraj Prakash Singh, Rajagopalan Srinivasan, Iftekhar Karimi

Digital Twin of Renewable Energy-Linked Power-To-Gas (P2G) Systems: Model Building and Continuous Update for Hydrogen Producing Electrolysis 18
Yongbeom Shin, En Sup Yoon, Dongil Shin

Keynote Talk - Artificial Intelligence and the Modern Workforce 19
Philippe Herve

On Demand Ceramic Open Cell for Intensified Reaction by High Precision Additive Manufacturing 20
Sang-Hoon Nam, Gian Han, Kavin Kowsari, Nicholas X. Fang, Seok Kim, Young Tae Cho

Additive Manufacturing of Hierarchical Porous Structures for Controlled Gas Bubble Flows Using Cellular Fluidics 21
Jonathan Davis, Anna Guell Izard, Hawi Gemeda, Erika Fong, Joshua R. DeOtte, Nikola Dudukovic, Sarah Baker, Eric B. Duoss

Cellular Fluidics: Tuning Multiphase Interfaces in 3D Using Architected Porous Media 22
Nikola Dudukovic, Erika Fong, Hawi Gemeda, Jonathan Davis, Sarah Baker, Eric B. Duoss

CO2 Capture Modeling with Intrastage Cooling 23
Joshua Thompson, Gyoung Gug Jang, Costas Tsouris

Thermal Energy Regulation with 3D Printed Polymer-Phase Change Material Composites 24
Peiran Wei, Ciera Cipriani, Emily Pentzer

Elucidating Metabolic and Regulatory Mechanisms of Microbial Aromatic Utilization for Lignin Valorization 40
Jinjin Diao, Tae Seok Moon

Depolymerization of Corn Stover Lignin with Bulk Molybdenum Carbide Catalysts 42
Xiaojun Yang, Maoqi Feng, Jae-Soon Choi, Harry M. Meyer III, Bin Yang

Developing a Multi-Target Alkali Sterilization Strategy to Facilitate Lignin Dispersion and Promote Biological Lignin Valorization 43
Zhi-Min Zhao, Shuyang Zhang, Xianzhi Meng, Yunqiao Pu, Arthur J. Ragauskas

Demethylated Lignin as the Interface Enhancer in FDM 3D Printing for PA12 Composites 44
Shuyang Zhang, Xianzhi Meng, Arthur J. Ragauskas

Biorenewable Chemical Production from Pyrolysate by Lactic Acid Bacteria 45
Samuel Rothstein, Swastik Sen, Thomas J. Mansell

Keynote Talk - Industry 4.0: Continuous Pharmaceutical Manufacturing Process 46
Ravendra Singh

Adaptive Strategies for Updating Unit Operation Models and in-Line Monitoring of Blend Uniformity in Continuous Pharmaceutical Manufacturing Process 48
Yingjie Chen, Shashwat Gupta, Andres Roman-Ospino, Fernando Muzzio, Marianthi Ierapetritou

Data-Driven Approaches Towards Equipment Health-Classification and Predictive Monitoring in Drug Product Manufacturing 51
Philipp Zuercher, Sara Badr, Hirokazu Sugiyama

An Equation-Oriented Model of a Trickle-Bed Reactor for Hydrodesulfurization Process Analysis and Digital Twin Applications 52
Heleno Bispo, Arianne Barros, Fernando V. Lima, Antônio Tavernard

Keynote Talk - Digital Transformation, Data Science, and Industrial Intelligence 54
S. Joe Qin

Keynote Talk - Decision Making in Digitalized Value Chain Optimization 55
Dimitri Papageorgiou

3D Stereolithography (SLA) Printing-Based Micro-Fabrication Using Custom Polymer Resin Chemistry for Rapid Prototyping of Microfluidic Chips and Component 56
Isteaque Ahmed, Katherine Sullivan, Aashish Priye

Mapping Process-Induced Chain Orientation in 3D Printed Parts 57
Anthony Kotula

Two-Wavelength Volumetric 3D Printing for Rapid Fabrication of Multi-Level Microfluidic Networks 58
Kaylee Smith, Sanaz Habibi, Martin de Beer, Zachary D. Pritchard, Timothy F. Scott, Mark A. Burns

Optical Interconnects on a Flexible Substrate Utilizing Additive Manufacturing Tools 60
Roger B. Tipton, Dianhao Hou, Thomas M. Weller, Venkat Bhethanabotla

Cold Spray of Polystyrene Particles on Various Substrates 61
Sebnem Ozbek, Michael J. Carter, Grant A. Crawford, Travis W. Walker

Use of Dimensionality Reduction and Transfer Learning in Deep Reinforcement Learning Controller for Hydraulic Fracturing 63
Mohammed Saad Faizan Bangi, Joseph Kwon

Harnessing Cognitive AI Technology in Refining to Enhance Operator Decisions 65
Leslie Rittenberg

Regression Model for Tool Wear Monitoring in Precision Machining 66
Seulki Han, Nasir Mannan, George Bollas

Selection of Combined Index Weights to Optimize Anomaly Detection in Big Area Additive Manufacturing 68
Monique McClain, Dhrubajit Chowdhury, Kris Villez

Keynote Talk - Topological Data Analysis: Concepts, Computation, and Applications in Manufacturing 71
Alexander Smith, Victor Zavala

Keynote Talk - Platforms and Algorithms for Digitally-Enabled Next-Gen Manufacturing 72
R Donald Bartusiak, Thomas Badgwell, John B. Vicente

Deep Composites: Bioinspired AI Towards Modeling, Design and Manufacturing 74
Markus J. Buehler

Evaluation of Process Variable Influence on Mechanical Properties of Short Fiber-Reinforced 3D-Printed Parts 75
Martin Etemadi, Arit Das, Michael Bortner

Direct Ink Write of High Solid Suspensions: Considerations in Particle Type and Binder Properties 76
Alexandra Marnot, Blair Brettmann

Dimensionalization of Two-Phase Newtonian/Non-Newtonian Flow Problems. 77
Abdul Salam Mohammad, Joseph J. Biernacki

Influence of Relative Humidity on the Spreadability and Triboelectric Properties of Powders in Additive Manufacturing Processes 78
Sebastien Depaifve, Aurelien Neveu, Filip Francqui, Geoffroy Lumay

Toward Next Level of Pharmaceutical 3D-Printing Through Advanced Lipid-Based Excipients 79
Moaaz Abdelhamid, MSc, Carolina Corzo, Martin Spoerk, Ioannis Koutsamanis, Carolina Alva, Ana Belén Ocampo, Eyke Slama, Dirk Lochmann, Sebastian Reyer, Sharareh Salar-Behzadi

Golem: A Probabilistic Approach to Robust Experiment and Process Optimization Based on Regression Trees 81
Matteo Aldeghi, Florian Häse, Riley J. Hickman, Isaac Tamblyn, Alán Aspuru-Guzik

Learning Based Scheduling of Industrial Hybrid Renewable Energy Systems 82
P S Pravin, Zhiyao Luo, Xiaonan Wang

Keynote Talk: Modelling and Monitoring with Dynamic Auto-Regressive Latent Variable Methods 84
Qinqin Zhu, Bo Xu, Haitian Zhang

Personalizing Wound Dressings Via 3D Printing for the Treatment of Thermal Burns.................................... 86
Jia Heng Teoh, Anbu Mozhi, Chi-Hwa Wang

3D Printing of Pharmaceuticals-Exploring Process Parameters and Structure Dry Weight Relations.............. 87
Mohammad Azad, Georgia Kimbell, Deborah Olawuni

Novel Polyester-Based Thermoplastic Elastomers for 3D-Printed Personalised Urethra Pessaries.................. 88
Martin Spörk, Florian Arbeiter, Ioannis Koutsamanis, Hrvoje Cajner, Matthias Katschnig, Simone Eder

3D Printing of Chiral Liquid Crystal Elastomers Using Cellulose Nanocrystals... 89
Mohsen Esmaeili, Kyle George, Nader Taheri-Qazvini, Monirosadat Sadati

Control Techniques for Handling Sensor and Actuator Cyberattacks on Evolving Nonlinear Process Systems.. 90
Henrique Oyama, Keshav Kasturi Rangan, Helen Durand

An Algorithm for Exascale-Capable Integrated Process Design and Control 92
Nikos Vasilas, Athanasios Papadopoulos, Lazaros Papadopoulos, Dimitrios Soudris, Panos Seferlis

Integrated Actuator Attack Detection and Control for Nonlinear Systems Under Lyapunov-Based Economic Model Predictive Control 95
Keshav Kasturi Rangan, Henrique Oyama, Helen Durand

Detectability-Based Controller Design Screening for Multiplicative Sensor-Controller Attacks 97
Shilpa Narasimhan, Nael El-Farra, Matthew Ellis

Expanding Resilient Lyapunov-Based Economic Model Predictive Control Concepts to a Distributed Control Framework.. 99
Dominc Messina, Henrique Oyama, Helen Durand

Commercialization of Aptamer-Based Electrochemical Sensors 101
Edgar D. Goluch

(Invited Talk) Developing Nanosensor Technology to Detect Cancer in Patients 102
Daniel Heller, Zvi Yaari

(Invited Talk) Wearable Health Sensors.. 103
Chelsea Monty-Bromer

(Invited Plenary Talk) Silicon Nanowire Photoelectric Protein Sensors 104
Marcie Black

(Invited Plenary Talk) Glucose Sensors for Bioprocessing: Progress and Perspectives.................................. 105
Kenneth Reardon

(Invited Plenary Talk) Optical Sensing and Cell Culture 106
Stephen Grant

Multiscale Modeling of Spray Coating of Perovskite QDs: Understanding the Role of Molecular Interactions in Particle Aggregation. 107
Niranjan Sitapure, Joseph Kwon

Findings and Conclusions from a Mobile Worker and Augmented Reality Enabled Continuous Manufacturing Skid Project.. 109
Iiro Esko, Katelyn Kelsey, Andrew Nachenberg, Leon Grossman

Practical Issues in Cybersecurity: From Encryption to Images 111
Dominc Messina, Kathryn Tyrrell, Minhazur Rahman, Kip Nieman, Keshav Kasturi Rangan, Henrique Oyama, Samantha Cherney, Arlan Bonislawski, Helen Durand

Keynote Talk: Machine Learning and AI Applications in the Chemical Industry 113
You Peng, Leo Chiang

Keynote Talk: Deploying AI for Automated Monitoring of Physical Infrastructure 114
Prateek Joshi

High Strength, High Toughness Parts Via Dual Material Fused Filament Fabrication 115
Brian Koker, Rebecca Ruckdashel, Hikma Abajorga, Ryan Dunn, David Kazmer, Eric D. Wetzel, Jay Park

Process-Structure-Property Relationships in Additively Manufactured Polypropylene Blends 116
Arit Das, Michael Bortner

Understanding Flow and Stress Development in 3D Printing by Material Extrusion 117
Bryan Vogt

Autonomic Self-Healing of 3D Printed Polymer Composites 118
Vinita Shinde, Asha-Dee Celestine, Lauren Beckingham, Bryan Beckingham

Synthesis, Characterization, and Application of Novel Surface-Eroding Photopolymer Formulations 119
Whytneigh Duffie, Kevin D. Barz, Tsvetanka S. Filipova, Timothy M. Brenza, Katrina J. Donovan, Travis W. Walker

Designing New Printable Thermoset Shape Memory Polymers Via Molecular Simulation and Machine Learning 120
Andrew Peters, Anwar Shafe, Aniruddha Chowdhury, Guoqiang Li, Collin D. Wick

Keynote Talk: Online Measurements for the Petrochemical Industry: Industry 4.0 Trends and Unmet Needs 121
Sherine George, James Tate, Paul Cammarata, Eric G. Schmidt, Rod Spitler, John Thibodeaux

Woodchip Moisture Content Estimation Using Short-Range Iot Wi-Fi for the Pulp & Paper Industry 122
Kerul Suthar, Jin Wang, Zhihua Jiang, Q. Peter He

Gradient-Weighted Class Activation Mapping (Grad-CAM) Based Explanations for Process Monitoring Results from Deep Neural Networks 124
Abhijit Bhakte, Bairi Sai Vasista, Rajagopalan Srinivasan

A Deep Learning Vision System for Classification of Manufacturing Defects 126
Christopher Hanselman, Asit Tiwari, Mamta Venugopal, Lingrui Cai, Yuanfang Guan, Edwin Comparini, Bo Zhang, William R. Prucka

Magnesium Oxychloride Formation Kinetics and Enhanced Water Stability for Sustainable Building Materials Applications 127
Christopher Kitchens, Saumye Vashishtha

Optimization of TEG Dehydration Process in Natural Gas Processing Under Metamodel Uncertainty 128
Rajib Mukherjee, Urmila Diwekar

Mapping Environmental and Economic Analysis of Decentralized Cogeneration Energy Management Centers 129
Nina Monteiro, Thomas Adams II

Development of an Interactive Software Tool for Designing Industrial Solvent Recovery Processes 131
Jake Stengel, John Chea, Emmanuel A Aboagye, Michael Mackley, James Geier, Kirti Yenkie

High Flux CO2 Selective Membranes for Renewable Natural Gas and CO2 Capture 134
Christine Parrish, Hannah Murnen, Sudip Majumdar, Ning Shangguan

Optimizing Energy Efficiency of Ammonia Production Via Electrochemical Reaction and Haber-Bosch Process 135
Gbemisola Ojo, Kyle Camarda

Teaching Process Data Analytics and Machine Learning 136
Richard D. Braatz, Weike Sun, Brian W. Anthony

Teaching Big Data Science to Undergraduate Students in the University of Iowa 137
Jun Wang, Joseph S. Gomes, Charles Stanier

Textile: Tutorials in Experimentalist Interactive Learning 138
Hawley Helmbrecht, Elizabeth Nance

Teaching Artificial Intelligence to Chemical Engineers: Experiences from a 35-Year-Old Course 141
Venkat Venkatasubramanian

Integrated Design and Model Predictive Control of Multiscale Systems Using a Multi-Fidelity Bayesian Optimization Approach 142
Farshud Sorourifar, Naitik A. Choksi, Joel Paulson

Adjustable Robust Optimization for the Planning Operations of Integrated Refinery-Petrochemical Site Under Demand Uncertainty 145
Zhang Lifeng, Zhihong Yuan, Bingzhen Chen

A Novel Hybrid Algorithm for Scheduling of Multipurpose Batch Process 146
Dan Li, Dongda Zhang, Nan Zhang, Jie Li, Liping Zhang, Xin Xiao

Impact of Split Delivery in Minimizing LNG Procurement Cost 148
Prashanth Ravula, Mohd Shahrukh, Rajagopalan Srinivasan, Iftekhar Karimi

Keynote Talk: Toward Autonomy for Safe and on-Demand Biomanufacturing on Mars 151
Ali Mesbah

Keynote Talk: The Roadmap to an Autonomous Chemistry Lab 152
Connor Coley

Multi-Objective Supply Chain Optimization in Personalized Healthcare 153
Andrea Bernardi, Niki Triantafyllou, Athanasios Antonakoudis, Matthew Lakelin, Nilay Shah, Maria Papathanasiou

Robust Bioprocessing of Lignocellulose Using Microbial Tipping Points 155
Katharine Hirl, Michael J. Shreve, John M. Regan, Tom L. Richard

Keynote: Multi-Objective Optimization, State Estimation, and Advanced Control of a Semi-Batch Process for the Enzymatic Conversion of Lactose into Value-Added Products 157
Ronald Alexander, San Dinh, Guilhermina Schultz, Marcelo P. A. Ribeiro, Fernando V. Lima

Keynote: Virtual-Engineering Software Framework for Integrated Biomass Conversion Modeling............... 159
Ethan Young, Hariswaran Sitaraman, Andrew Glaws, Andrew Bartling, James J. Lischeske, Jonathan Stickel

Comprehensive Planning of Annual Delivery Program for LNG Suppliers.. 160
Dnyanesh Deshpande, Mohd Shahrukh, Rajagopalan Srinivasan, Iftekhar Karimi

A New Integrated Scheduling and Optimization Framework for Holistic Refinery Supply Chain Management.. 162
Li Yu, Qiang Xu

Integrated Load Shifting and Curtailment for Demand Response of Central Chilled Water Plants................ 163
Gustavo Campos, Yu Liu, Ahmet Palazoglu, Nael El-Farra

A Web-Based Application for Chemical Production Scheduling.. 164
Venkatachalam Avadiappan, Lucas Buttazoni, Shamik Misra, Hojae Lee, Martin Yang, Christos Maravelias

Keynote Talk - Positive Power with Negative Emissions: Flexible Ngcc Enabled by Modular Direct Air Capture (DAC)... 165
Matthew Realff, Fani Boukouvala, Christopher W. Jones, Fanhe Kong, Ryan Lively, David Thierry, Joseph K. Scott, Howard Hendrix, Katherine Dombrowski, Darshan J Sachde, Andrew Sexton

Keynote Talk - Machine Learning of Molecular and Materials Properties at the Low-Data Limit................. 166
Srinivas Rangarajan, Huijie Tian, Bowen Li

Model-Based Investigation of Upstream CHO Cell Culture Process for Production of Monoclonal Antibodies with Desired N-Linked Glycosylation 167
Ou Yang, Jayanth Venkatarama Reddy, Katherine Raudenbush, Aron Gyorgypal, Shishir Chundawat, Marianthi Ierapetritou

Next-Generation Vaccines and Therapeutics: Towards Resilient Pharmaceutical Supply Chains.................. 169
Miriam Sarkis, Nilay Shah, Maria Papathanasiou

Systematic Decomposition & Evaluation of a Process Design Space for Monoclonal Antibody (mAb) Manufacturing.. 171
Johann Kaiser, Maria-Ona Bertran, Janus Krarup, Manuel Pinelo, Ulrich Krühne, Deenesh K. Babi

Keynote Talk: Integrated Quality by Design in (Bio)Pharmaceutical Manufacturing 172
Richard D. Braatz, Moo Sun Hong, Amos E. Lu, Weike Sun

Keynote Talk: An in Silico Approach for Monoclonal Antibody (mAb) Process Research & Early Development 173
Johann Kaiser, Maria-Ona Bertran, Janus Krarup, Manuel Pinelo, Ulrich Krühne, Deenesh K. Babi

Developing a Workflow for Continuous Centrifugal Extraction... 175
Eric Moschetta, Benjamin Rizkin

A Scalable Continuous Reaction and Isolation Process for the Production of Sulfonyl Chloride Pharmaceutical Intermediates.. 176
Matthew Glace, Cameron Armstrong, Michael Scott, Thomas Roper

Target Polymorphic Form Development Via Continuous Combined Cooling and Antisolvent Crystallization Using Oscillatory Baffled Crystallizer 178
Shivani Kshirsagar, Naga Lakshmi Ramana Susarla, Srividya Ramakrishnan, Zoltan Nagy

Development of a Novel Continuous Spatially Distributed Diafiltration Unit Operation 180
Xiaoyan Long, Zoheb Khan, Eoin Casey, Denis Dowling, Steven Ferguson

Predicting Screw Feeder Flow Rates from Powder Properties and Operating Conditions 182
Brad Johnson, Bram Bekaert, Valérie Vanhoorne, Thomas De Beer, Salvador Garcia-Munoz, Nikolaos Sahinidis

Model-Based Analysis of Breakage in Fluid Bed Drying of Continuously Produced Pharmaceutical Wet Granules 184
Michael Ghijs, Tuur Vandeputte, Marie Vandromme, Selien Van Langenhove, Thomas De Beer, Ingmar Nopens

Development of a DEM-Based Digital Twin of a Continuous Direct Compression Line 186
Peter Toson, Pankaj Doshi, Peter Bohling, Martina Trogrlic, MSc, Marko Matic, Daniel O. Blackwood, Kai Lee, Marta Moreno Benito, James Kimber, Hugh Verrier, Johannes G. Khinast, Dalibor Jajcevic

Physical Property Characterization of Ionic Liquids in HFC-32 and HFC-125 Using a High-Pressure View Cell 187
Greta Olsen, Kalin Baca, Mark Shiflett

Material Selection and Design of Ionic Liquid-Based Extractive Distillation for Hydrofluorocarbons Separation 188
Mohammed Sadaf Monjur, Ashfaq Iftakher, M M Faruque Hasan

The Effect of Alkyl Chain Length on Imidazolium-Based Ionic Liquids and Choice of Halide Anions (Cl, Br, I) on Phase Equilibria for HFC-32, HFC-125, and Binary Mixtures (R-410A) 190
Kalin Baca, Greta M. Olsen, Lucia Matamoros Valenciano, Madelyn G. Bennett, Mark Shiflett

Liquid-Liquid Equilibria of Ternary Systems of Thermally Robust Ionic Liquids and Hydrocarbon Mixtures 191
Santosh Rathan Paul Bandlamudi, Samuel Todd, James H. Davis Jr., Kevin West, Brooks Rabideau

The Deterministic Role of Hydrogen Bonding on Water Diffusivity in Methylimidazolium Halide Ionic Liquids and Ionogels 192
Alexandra V. Bayles, Julia Fisher, Connor S. Valentine, Arash Nowbahar, Matthew Helgeson, Todd Squires

Applications of Phytoremediation for Catalytic Enhancement of Biocarbon Towards Supercapacitors 193
Katelyn Shell, Shaan Vohra, Dylan Rodene, Ram Gupta

Electrochemical Studies on Graphitized Biocarbon Derived from Hydrothermally Liquefied Low Ash Content Corn Stover 194
Katelyn Shell, Vinod Amar, Sergio Hernandez, Rajesh Shende, Ram Gupta

Electrocatalyst Development for Making Ammonia from Air and Water at Ambient Condition 195
Jian Liu, Xiaohong Xie, Yuyan Shao, Manh-Thuong Nguyen, Vassiliki-Alexandra Glezakou, Daniel Z. Deng, Robert Cavagnaro

Ammonium Formate as a Safe, Energy-Dense Electrochemical Fuel Ionic Liquid 196
Zachary Schiffer, Karthish Manthiram

Enhanced CO2 Electroreduction to CH4 and C2H4 Via Selective Proton Transfer 197
Marcel Schreier

Bridging the Knowledge Gap by Empowering Engineers with IIoT Advanced Analytics 198
Vincent Teran

Biodirected Assembly of Ionomer Thin Films for Fuel Cells and Electrolyzers 199
Julie N. Renner

Parametric Testing, Analysis and Techno-Economic Assessment of 2-Eempa, a Single-Component Water-Lean Post-Combustion Capture Solvent 200
David Heldebrant

Diversification of Basic Immobilized Amine Sorbents (BIAS) for Carbon Dioxide Capture 201
McMahan L. Gray, Qiuming Wang, Walter C. Wilfong, Shouliang Yi, Tuo Ji, Fan Shi

CO2 Capture Modeling, Energy Savings, and Heat Pump Integration 202
Y. Liu

Process Intensification for Direct Conversion of Biomass-Based Syngas to High Octane Gasoline 203
Claire Nimlos, Daniel Ruddy, Connor P. Nash, Dan Dupuis, Anh To

Carbon Oxide-Thermic Oxidation Process for Synthesis of Porous Silicon-Carbon Composite Anodes for Lithium-Ion Batteries 204
Mohamed Ali Elharati, Younghwan Cha, Qusay Bkour, Steven Saunders, M. Grant Norton, Min-Kyu Song

Optimal Design of Controlled Environment Agricultural Systems Under Market Uncertainty 205
Shaylin Cetegen, Matthew Stuber

Reproducible and Fast Mapping of Material Flows in Industrial Networks Using Piot Hub: A Novel Cloud Based Computational Tool 208
William Farlessyost, Apoorva Bademi, Shweta Singh

Solid Oxide Fuel Cell – Thermal Energy Storage (SOFC-TES) Integrated System Modeling and Energy Efficiency Optimization. 209
DongJin Lee, Youngtak Cho, Sungwon Hwang

Upcycling Food Waste for Hydrogen Storage: Technoeconomic Assessment 211
Al Ibtida Sultana, Toufiq Reza

Economic and Environmental Impact of Biochar Addition on Anaerobic Digestion Process 212
Md Mosleh Uddin, Zhiyou Wen, Mark Mba Wright

Optimizing Syngas Production Process by Downdraft Gasifier Using Response Surface Methodology 213
ASO Hassan, Hayder Alhameedi, Salih Rushdi, Hayder Al-Atabi, Zainab Al-Sharify, Joseph D. Smith

Techno-Economic Evaluation of Nano-Porous Silica Production from Rice Husk and Sand 214
Semie Kim, Young-Il Lim

A Capital and Energy Efficient Alternative Emissions Control Technology for Controlling Emissions Using Bead Activated Carbon 215
Kim Tutin, Harold Cowles, John Berger

Conversion of Plastic Waste to Fuel: Assessment of Technologies and Economics 235
Cesar Lubongo, Taylor Congdon, Paschalis Alexandridis

Continuous Back to Monomer Recycling of PET from Composite Materials Using a Twin-Screw Extruder 236
Esther Brepohl, Lars Biermann, Mandy Paschetag, Hannes Schneider, Carsten Eichert, Stephan Scholl

The Effect of Residence Time and Heterogeneous Catalysis on Continuous Hydrothermal Liquefaction of Food Waste 238
Aristidis Mihalos, Geoffrey Tompsett, Michael T. Timko, Alex D. Paulsen

Continuous Hydrotreatment of Sewage Sludge and Spirulina Algae Biocrudes from Hydrothermal Liquefaction: Different Catalysts for Different Organic Contaminants 239
Muhammad Salman Haider, Daniele Castello, Lasse Rosendahl

Review and Modeling of Organic Removal in Anaerobic Sludge Blanket Bioreactors for Methane Production and Energy Recovery from Domestic Wastewater 240
Chelsea Q Linvill, Ian Morris, Mia Padon, Andrew Pfluger

Utilization of Marble Dust as a Concrete Additive to Enhance Its Properties 241
Sejal Ahuja, Dr. Manpreet Singh, Dr. Utkarsh Maheshwari

Deployment of Machine Learning Models in Pharmaceutical Development 242
Jose Tabora, Patrick Sipple

Unsupervised Learning from Sets of Data Using Contrastive Latent Variable Models 243
Kristen Severson

Energy Dispersive X-Ray Hyperspectral Image Analysis and Chemometrics for Catalyst Characterization 244
Jose Maria Gonzalez Martinez, Jose Manuel Prats Montalban, R. Haswell, Alberto Ferrer

Applications of Machine Learning in Production Enhancement and Field Monitoring in Oil & Gas Industry 245
Haijing Gao, Shuxing Cheng

Author Index

630151 Influence of Interfaces in Electrical Properties of 3D Printed Structures

Abstract:

Additive Manufacturing technology has advanced to being a major player in the field of manufacturing in the last two decades revolutionizing the manufacturing of mechanical parts for aerospace and automotive applications, prosthetics, and dentistry. The application of additive manufacturing in the production of sensors and actuators can alleviate the delivery and storage related constraints for remote locations such as offshore platforms and space stations. Material extrusion-based additive manufacturing (MEAM) is a uniquely suited additive manufacturing technique for cost-effective printing of sensors and actuators. In MEAM, a 3D structure is built in a layer-by-layer manner as instructed by a g-code, by extruding a thermoplastic material through a nozzle heated above the melting point of the plastic. Several parameters including extrusion temperature, print speed, bed temperature, part orientation, layer height, raster angle, and infill can be varied in MEAM. Contemporary simulation and experimental studies, both show that the print parameters play an important role in determining the strength of interlayer bonding and the resulting anisotropy in mechanical strength of printed parts. To develop MEAM into an industrially feasible technique for 3D printing sensor and actuator components, similar fundamental studies are necessary to predict electrical part performance based on print parameters.

In this study, we conduct work using a commercial conductive composite of polylactic acid and carbon black. We adopt a hollow box structure specimen to examine the effect of extrusion temperature and print speed on the electrical characteristics of the bond interfaces, specifically those between two subsequent layers. The walls of the hollow box compose of single extruded fibers stacked on top of each other. The electrical characteristics are measured using small signal impedance spectroscopy to avoid Joule heating effects. The impact of varying the extrusion temperature and print speed on electrical impedance was examined in two directions, across the interlayer bond interfaces (Z-direction), and along the interlayer bond interfaces (F-direction). A variation of impedance in F-direction with extrusion temperature was observed and not print speed, indicating that the material resistivity varied with extrusion temperature. Thus, the Z/F ratio was used to quantify anisotropy in electrical impedance as the extrusion temperature and print speed were varied. It was found that the Z/F ratio remains relatively constant with variation in extrusion temperature and print speed. Variation in sample cross-section in Z- and F- direction were found to result in a Z/F ratio of 1 in COMSOL Multiphysics simulations. This showed that the bond interfaces were primarily responsible for the observed Z/F of 2.15 ± 0.23, that is, the bond interfaces contribute nearly two-third of the impedance when measuring across the fibers.

To understand how the impedance scales with the number of interfaces, impedance was measured across different number of bond interfaces in samples printed with extreme parameters - high/low print speed and high/low extrusion temperature. Impedance was found to scale linearly with the number of interfaces. This implied that tests across single bond interfaces could be used to predict results over multiple interfaces; thus, allowing to study effects of multiple parameters in a relatively short time. The scaling law was found to change when using different extrusion temperature; high extrusion temperature has a lower impedance per interface.

Thus by investigating the electrical behavior of bond interfaces, we provide a framework to model and predict anisotropy in electrical impedance and predict electrical impedance based on print parameters.

630223 Modeling Heat Transfer in Material Extrusion Additive Manufacturing: Balancing Model Resolution with Computational Demands

Abstract:

Thermal modeling informs the understanding of cooling phenomena and structural evolution during material extrusion (MatEx) additive manufacturing to help predict and tune final part properties. For fused filament fabrication (FFF) and big area additive manufacturing (BAAM), in particular, thermal models are well-suited to augment parameter optimization and closed-loop control for high throughput manufacturing of custom parts and products. However, limitations in computational time, speed, and cost pose a challenge to modeling large structures with sufficient multi-dimensional granularity over extended timescales, with many models confined either to high resolution studies on short length and timescales or low-resolution, lumped approximations for larger simulations. Thus, the goal of this work is to develop a high fidelity thermal model applicable from desktop to BAAM systems with reduced computational constraint.

We present a scalable, two-dimensional finite volume model and algorithm to rapidly simulate thermal profiles across both FFF and BAAM length and time scales. Our approach implements a two-dimensional implicit numerical strategy that confers reduced computational complexity while offering fine scale granularity in capturing critical inter- and intralayer cooling dynamics and radiative heat transfer. Further, analysis of the effective Biot number (*Bi*) reveals that, on the BAAM scale, layer width and radiative heat transfer both drive the formation of intralayer temperature gradients, exceeding 51°C between the layer center and edge. We further simulate the fabrication of single road-width acrylonitrile-butadiene-styrene (ABS) walls at the BAAM scale (50 layers) to demonstrate the relative speed of the computational approach and its ability to capture the complex dynamics of a multilayer build. The model agrees with published experimental measurements, and results gauge the influence of layer thickness and proximity to print bed on cooling behavior, with larger layers and separation from the print bed lending to longer periods for interlayer welds to form above the ABS glass transition temperature (T_g). Finally, a BAAM case study highlights the model's utility in design for additive manufacturing (DfAM) and print optimization, demonstrating how changes in bed temperature and layer time can impact interlayer bonding.

631340 Assessing the Fidelity of Additively Manufactured Objects

Abstract:

Despite the rapid growth of additive manufacturing (AM), its broad implementation is still hindered by challenges including limited material options, inconsistent part quality, and fabrication defects. Although current additive manufacturing technologies use high-precision machines, in practice, manufactured parts can deviate significantly in both shape and structure. Thus, characterization of the fidelity of manufactured objects is critical to establish and ensure performance. Although there are many ways to compare printed objects and the digital counterparts from which they are created, many of these methods are biased due to the way that measurements are made or the subjectivity of alignment procedures. To remove these subjectivities and biases, alternative printability metrics were developed based on a new alignment technique. Rather than using subjectively selected landmarks, the new method uses the axis of the minimum moment of inertia as reference. A boundary-based printability index (PI_B) and a volume-based printability index (PI_v) were then used to assess external dimensions and object volume. X-ray computed tomography (X-CT) images of 3D printed objects were used. Guidelines for statistically sound sampling were also developed. The results show the extent to which different metrics are sensitivity to different types of flaws and offer ways of comparing the effect of processing conditions on print fidelity.

631938 Design and Processing of Open Lattice Structures for Tunable Fluid Phenomena

Abstract:

The rapid evolution of additive manufacturing has realized previously unproducible designs, such as intricate lattice networks, which have demonstrated utility in heat exchangers, reaction vessels, structural scaffolds, biomedical implants, and even consumer products. A growing body of research has considered this class of materials for applications in process intensification, thanks to the variety of available designs and the range of available dimensions. However, much of the basis for these investigations is founded on historical use of stochastic open cell foams, and thus the potential of additive manufacturing for designing purposely tailored structures for fluid applications remains largely untapped. Furthermore, process variability further contributes hindrances to the adoption of these technologies. Here we investigate an assortment of lattice architectures in the context of fluid dynamics applications. Using the production-oriented Carbon M1 3D printer, we produce lattice parts for experimental pressure studies with homogeneous and heterogeneous unit cell configurations. Upon dimensional characterization, we observe that the photopolymerization process results in a scale-dependent gradient deviation from as-designed part dimensions, to which fluid processes may be more sensitive than structural applications. In response, we demonstrate a process for design-stage compensation to improve agreement between model predictions and experimental results. We further examine how scaling of the unit cell and overall part geometry contribute to fluid phenomena within these ordered open cellular structures and present guidelines for designing lattice configurations. With this complete pipeline, we illustrate the potential for additive manufacturing to produce tailor-made, multifunctional process components for chemical engineering applications.

621938 Methods to Evaluate Residual Stress in FDM Printed Parts

Abstract:

Additive manufacturing has gained a lot of attention lately because of its ability to develop prototypes directly from a CAD model. However, due to the layer-by-layer sequential solidification of the printed material and large temperature differences that result during the printing process, residual stresses are prone to develop in AM parts, leading to geometrical inaccuracies like volume shrinkage, warpage and distortion of the printed part, adversely affecting the final product quality. Measuring and controlling such residual stresses can therefore help us better choose the printing parameters and hence greatly enhance the dimensional accuracy of the printed parts. In this study, Fused deposition Modeling (FDM) is employed to study the effect of printing speed, printing temperature and nozzle diameter on the residual stress generated in commonly used FDM filaments like ABS and PC printed parts. Volume shrinkage of ABS printed samples is quantified suing Dynamic Mechanical Analysis (DMA). The destructive hole-drilling technique coupled with a 2D subset-based digital image correlation (DIC) software is used to measure the residual strains. Non-destructive optical residual stress measuring techniques like RAMAN spectroscopy and birefringence are also used for characterization. Finally, an experimentally validated finite element analysis (FEA) model is developed to simulate the FDM printing process and the hole-drilling method. A non-linear temperature-dependent viscoelasticity model is also developed to predict the behavior change of the filament as it solidifies. The simulations and models developed can be directly applied to predict the residual stresses that are generated during the FDM printing process and hence fine tune the printing parameters leading to an improved final product quality.

ACCELERATING TIME TO AUTONOMY IN AN INDUSTRY 5.0 WORLD

Brent Kedzierski, HumanWRKS LLC, Houston, TX
Ian Willetts, AVEVA Software LLC, Carlsbad, CA

Introduction

Industry 4.0 is about smart production, IoT, sensors, drones and other cyber physical innovations. Industry 5.0 will be about connecting smart production to human cognition, collaboration, and creativity. To this end, progress will be judged on not just the connection of data between the Internet of Things (IoT), but how that connectivity enhances the human experience and shortens the time to achieve real autonomy for today's connected worker.

This paper discusses how AVEVA are working together with international oil companies (IOCs) to develop and deploy more personalized worker experiences to serve the new industrial agenda of achieving the goals of both, the enterprise as well as continually ensuring the welfare of humans.

Industry 4.0 Needs People with Connected Worker Skills

Industry 4.0 emerged in 2011 and has been the most technologically sophisticated and focused of all industrial revolutions. Its focus has been on transforming industrial manufacturing through digitalization and the exploitation of the potentials of new technologies. The suite of connected technologies that have been adopted across industry over the last decade are commonly referred to as the 9 pillars of Industry 4.0[1] (Simulation, Augmented Reality, Big Data, Additive Manufacturing, Cloud Computing, Autonomous Robots, Cybersecurity, IIOT, Horizontal and Vertical System Integration). These pillars seek to work together as disruptive connectable technologies to transform industrial production by creating an industrial ecosystem that is both automated and intelligent.

However, research reveals that many manufacturing leaders believe there are a lack of skilled workers across industry to support their digital transformation goals[2]. The skills gap is understandable given that confidence and adoption levels of automation and digital transformation innovations are typically spread across a bell curve of innovators, early adopters, early majority, late majority, and laggards. The second decade of Industry 4.0 will see mature connected ecosystems become common place across major industries. However, the lack of requisite connected workers will have a dual effect on the advancement of smart industries:

- First, industry will experience a value degradation in their implementations of advanced technologies without enough autonomous and digitally fluent workers.
- Second, the industrial workforce will become increasingly disengaged as opportunities to grow their skills are superseded by a focus on automation, resulting in issues related to workforce attraction, retention and development.

Workforce competence development in the past has always been understood to be a critical enabler of operational excellence and safe operations. However, as connected technologies mature industry must take a long hard look at developing a highly competent and autonomous workforce in parallel to the deployment of their Industry 4.0 initiatives.

Shift from Competence to Autonomy

Across industry the phrases "time to competence" and time to autonomy" have often been used interchangeably. However, autonomy is a very individual state of being that goes beyond an organization simply giving an employee the license to act based on the assurance of their competence.

To provide context on the difference between competence and autonomy we can consider a driver that is learning to drive a car. They will study the theory and take driving lessons on the road with a view to passing the driving test. The driver will reach a level of competence so they can pass the test and know "how" to drive a car. However, that driver will only become truly autonomous after an extended period of driving where they are comfortable in the seat and able to be aware of their surroundings, understand the cause and effects of their decisions and truly contribute to a safe driving experience for everyone.

Even when world class organizations such as Shell took a concerted effort to implement a global development program in 2011 the program was based on a traditional programmatic competence development and assurance program and not designed to enable true worker autonomy. The program was effective in reducing time to competence requirements, but competence alone does not equal autonomy. At best competence provides management with a single factor level of assurance that workers possess the requisite knowledge and skills linked to a standardized framework.

To succeed in today's tech heavy industrial settings and meet evolving worker expectations, industry must move past the traditional view of competence to drive towards a true level autonomy across its workforce. The advanced technologies of Industry 4.0 will require a more human centric framework for worker autonomy in order to truly enable the optimization of the human machine mix.

The actual tools of the Industry 4.0 revolution such as the increased availability of real time data, simulation, cloud computing and virtual reality can be leveraged to not only improve the manufacturing process, but they can also be used to accelerate worker engagement and time to autonomy for the connected worker.

Industrial Workforce Autonomy – Industry 5.0

There have always been six (6) essential components of workforce autonomy. Each one of these components fall under the three (3) core areas of human ability:

- Cognition
- Creativity
- Social Connectivity

The desired end state is ensuring the worker can consistently:

1. Execute job tasks, analyze, synthesize, and evaluate relevant job/task data
2. Apply sharp cognitive abilities to apply data to troubleshoot and make decisions

3. Apply their creative abilities to solve problems
4. Leverage their social skills to collaborate with colleagues in all areas of work
5. Know when and when not to escalate issues.

Looking at each of the above components of autonomy, industry must shift from a technology centric 4.0 viewpoint to a connected worker 5.0 viewpoint. Industry 5.0 will represent a human centric revolution where technology will play a secondary role in the hunt for industrial expansion.

The modern industrial worker must be equipped with the skills to thrive and work autonomously with intelligent technologies and data rich cultures to drive advances in innovation and sustainability. Industry 5.0 has entered the scene because industry, society and the types of gains required in today's economy are at a place where a value-based revolution is needed. Over the next decade industry will look within every nook and cranny of the industrial framework to reposition the human and the human-machine mix to deliver a new level of value for industrial stakeholders.

Strategies for Building Autonomy

The need for a more strategic development model for workers is now at a critical tipping point across industry. These strategies must work to develop humans that can not only co-exist but thrive in a tech heavy industrial world.

The jobs that are emerging due to connected technologies will have higher order intellectual, social, creative, and behavioral components that will present new challenges to human competence, efficacy and autonomy. Part of the new upskilling and reskilling dilemma is that traditional industrial skills must shift to match the demands of future connected industry. The opportunities across the entire industrial ecosystem to evolve new skills that derive autonomy can't be developed by osmosis or with subpar learning experiences as in previous decades.

In any effort to transform an organization, the first step is to establish a sense of urgency based on recognizing the need to change[3]. Also, complex, highly automated production systems leave fewer opportunities for on-the-job training, while also raising the stakes of human error in abnormal situations. These changing circumstances create an urgent need to improve training, but industry is slow to shift from traditional approaches. We also must recognize that today's workforce has very different expectations about technology which is important because technology can underpin new ways to drive toward autonomy.

From Training to Learning in the Flow

In their book, Minds at Work: Managing for success in the knowledge economy, David Grebow and Stephen J. Gill explain that organizations will need to shift from a training culture to a learning culture[4]. Training is the giving of information and knowledge, through speech, the written word or other methods of demonstration in a manner that instructs the trainee. Learning is the process of absorbing that information to increase skills and abilities and make use of it under a variety of contexts.

Learning is aligned with results. Learning-centered organizations make information accessible and individually relevant. Competence is judged by one's demonstrated abilities, not by an accumulation of seat time. Continuous learning in areas that matter to worker identity is key. People constantly create, share, and adopt information to make the company smarter in a learning organization.

Simulation Based Experiential Learning

Learning by doing has been a key mantra in the learning field for decades. Good simulation-based learning designs have been based on ensuring a few basic instructional practices are embedded into learning design. These included elements such as ensuring a clear goal, providing relevant context, challenge, giving learners choice options, showing consequence of actions, and coaching or feedback support. According to the NTL Institute for Applied Behavioral Science[5], simulation-based experiential learning produces more than seven times the knowledge retention of other training methods5.

In building autonomy, the goal is to build worker efficacy in their freedom to make choices for themselves rather than relying on others. Simulation based learning provides a unique learning experience where the decisions made by the learner on a simulator, during a scenario-based exercise, creates a realistic feedback loop where the consequences of an action, good or bad is seen and understood. The context rich nature of simulation also creates a unique learning experience for every single person, every lesson. There is no better way to build autonomy than by seeing and learning from the consequences of your actions in real time.

Immersive Training Simulators (ITS)

Immersive simulation tools use augmented reality and virtual reality to connect control room operators, maintenance, and field personnel in a single realistic learning environment. It is broadly used for site familiarization, work policies, field operations training and maintenance procedures. Augmented reality can be especially useful for remote collaboration with experts.

While all people are different and learn in different ways, simulation-based learning can help to cement lessons most effectively, using an immersive yet safe replication of real-world scenarios. Rather than simply knowing what they should do, immersive training allows users to see, hear, and feel what to do, so that when they are deployed in the real world, the right habits will have already been ingrained. One of the main principles of a learning-centered organization is to encourage measured risk-taking and experimentation. In an industrial environment, simulation is often the only safe, practical option for experimenting with risk-taking.

AVEVA has delivered to customers a full complement of behavioral learning modules utilizing the Cloud hosted AVEVA XR Virtual Reality platform.

Operator Training Simulators (OTS)

Operator error is a well-known problem. In the US alone, more than 84 incidents occur per week[6]. Chemical plants lose more than $20 billion annually from human error and insufficient knowledge[7]. A poorly trained operator can undo the most advanced engineering, placing a company's people and profit at risk.

An operator training simulator combines dynamic process simulation models with emulated control systems to create a virtual control room specific to a single plant or process. An operator training simulator is a building block of Shell's Multi-Purpose Dynamic Simulation (MPDS) initiative which uses the dynamic process model to support FEED and detailed design, the integrated dynamic model and controls database for virtual controls commissioning, establishing operating procedures and for operator training that provides management with confidence that that operators are ready for day one.

Eric Scappazzoni of the TOTAL Energies Manufacturing & Projects Division when interviewed said that he strongly believes using OTS is really improving TOTAL's operational excellence by increasing the competency and reducing the risk of having accidents. Other IOCs are currently executing programmatic deployments of Operator Training Simulators with AVEVA with a view to driving operational excellence and increased safety, while upgrading their learning experience for operators.

Historically OTS systems have been delivered on premise with a relatively large hardware footprint. Companies have realized the expected benefits of such systems in the early deployment period; however, a common challenge is that companies are often not willing to invest in long-term maintenance. AVEVA has addressed the challenge of the long-term maintenance of on-premise systems by providing the option to deploy both OTS and ITS systems to the Cloud.

Cloud Enabled – Anywhere, Anytime, Any Device

The optimum learning platform must leverage the Cloud pillar of Industry 4.0 to empower the connected worker to be autonomous in choosing when and where they invest their time and attention for their individual development. Today's simulators are more advanced than ever and bolstered by Cloud delivery that eliminates many of the traditional barriers to incorporating simulation-based learning into the curricula and serving the content up to the workforce anytime, anywhere and on any device.

This last point is paramount as the competition for learner's time and attention grows, the relevant factor will be where learners see the investment their time and attention paying the most dividend. Choice is about where the learner feels they can benefit most and in line with autonomy that means providing workers with greater control over both their work and their own learning and development.

Instructional Design

As the world of work and learning continue to blend closer together, it is important to consider models that factor in more human centric theories that are not only based on good learning design but are also relevant to driving autonomy. Stronger technology-based learning programs are much better equipped to incorporate higher concept and higher touch design elements that accelerate autonomy. These include considerations from human centric theories such as self-determination, self-efficacy, expectation value assurance theory, as well as attributes from human fulfillment and flow. Learning designs that factor in the elements of these theories will help shift industry from simply building competence against a static portfolio of skills to driving autonomy across a dynamic portfolio of worker enabled outcomes.

With the above platform that leverages three of the nine pillars underpinning Industry 4.0, organizational learning capabilities will mature to create content and context rich scenarios that will be less rote but instead more dynamic and challenging. Learning capabilities will mature to offer multi-model content that provides the learners:
More choices during each learning exercise

- The ability to analyze tasks and synthesize and evaluate relevant information
- The opportunity to evaluate conditions, make decisions and see "what if" consequences
- The chance to identify opportunities for hierarchical and matrixed collaboration

The learner is no longer watching video content and checking boxes. Instead, the worker is on a path to renewal and autonomy and all the benefits therein.

Unified Learning

Companies already invest substantial time and money into operator training. Simply investing more may yield some improvement, but it will not generate the transformation that the problem demands. In the Baker Report on the Texas City disaster, investigators noted that BP's computer-based training produced little retention or competence among operators, because the culture treated learning as just a box to check and then move on[8].

AVEVA's Unified Learning platform moves companies beyond point in time solutions such a rote e-Learning content such as that previously deployed by BP to a fully integrated Cloud hosted learning platform including:

- Instructional System Design methodologies
- E-Learning courses
- Process Competency Modules
- Generic Training simulators
- Custom High Fidelity Training Simulators
- Immersive Training Simulators
- AI Infused Microlearning
- Learning Management System

Rather than using disparate point solutions AVEVA's Unified Learning provides the optimum environment to allow companies to accelerate the time to autonomy for their workforce during this Industry 4.0 industrial revolution.

Conclusion

Without building more direct paths to higher levels of individual autonomy, industry will lack the ability, flexibility, and adaptability to optimize the connected technologies of Industry 4.0 in ways that fully harness and integrate the full spectrum of unique human capabilities.

Great companies will build human performance experience models that enable getting work done and harvest a core byproduct with opportunities for the worker to experience greater levels personal fulfillment and growth.

AVEVA is working diligently to ensure that its Unified Learning Platform provides all the necessary infrastructure and thought leadership to accelerate the time to autonomy for the workforce of today and tomorrow.

References

1. Wai Yie Leong, ; Joon Huang Chuah, ; Boon Tuan Tee, (ed.): 'The Nine Pillars of Technologies for Industry 4.0' (Telecommunications, 2020) DOI: IET Digital Library, https://digital-library.theiet.org/content/books/te/pbte088e
2. Hipplod, Sarah. Gartner Press Release. STAMFORD, Conn., May 11, 2021.
3. Kotter, J. P. (1995). "Leading Change: Why Tranformation Efforts Fail," Harvard Business Review.
4. Gill, S. J., & Grebow, D. (2017), "Minds at Work: Managing for success in the knowledge economy," Alexandria, VA: ATD Press
5. NTL Institute for Applied Behavioral Science
6. ARC Advisory Group
7. Abnormal Situation Management Consortium (2019)
8. Baker, J. A. et al (2007) "The Report of the BP U.S. refineries independent safety review panel"

622059 How to Get a Head Start on Your Digitalization Journey

Abstract:

Companies of all types and from all industries are increasing their productivity with digital technologies. With the right tools, they have the ability to increase efficiency, security, and productivity. They can achieve this by taking a holistic view of existing strategies and identifying smaller, more manageable process optimization opportunities of the existing installed base. In doing this, they need to leverage technologies such as artificial intelligence, smart monitoring and data analytics. The outcome is a customized digital transformation roadmap.

It is relatively easy to talk about the need for a digital transformation from a high level, but when it comes to planning and executing the strategy to create the specific roadmap, it tends to be more challenging. But it doesn't have to be that way. Industry 4.0 has now been around for a few years, which has allowed tried and tested approaches and methodologies to emerge that can be applied to pretty much any size company in any industry.

During this presentation you will gain the following insights:

- How to take the first steps of the digital transformation journey
- View of simulated business cases to dive deeper into the basic understanding of how to identify relevant, Industry 4.0 projects
- How to best prioritize these in terms of resources, return on Investment, digital maturity level, and more.

632003 Development of an Advanced Control Infrastructure for Subcritical Power Plant Online Power Demand Tracking

Abstract:

Many subcritical coal-fired power plants that usually operate in baseload conditions are migrating to load-following operations as more renewables are penetrating into the power grid. As renewables, such as solar and wind, deliver intermittent power, dispatchable energy systems such as coal-fired power plants are required to cycle their load to accommodate power demand fluctuations. However, most of the subcritical plants were not designed to operate under load-following conditions, resulting in additional emissions and increased power plant damage. Given that cycling may become the new operating mode for coal-fired power plants, there is a critical need to properly control these plants. In particular, to operate in different power dispatching modes, a power plant will need to produce steam at certain flows, temperatures and pressures as required by the turbine. The highly variable resulting plant dynamics will require predictive control algorithms capable of maintaining the process within the operating regions of interest.

In this work, an advanced control structure is developed for online control of the total power demand of a subcritical coal-fired power plant cycling around the setpoint of 200 MW. This structure involves the combination of proportional-integral-derivative (PID) and advanced model predictive control (MPC) algorithms [1] for real-time control of power demand. Specifically, temperatures of the primary and secondary desuperheaters are controlled by a 3 input x 3 output MPC that manipulates 2 spray valves and coal flowrate as inputs. Also, the PIDs (2) control the pressure and water level of the drum boiler by manipulating the water valve and steam-to-turbine valve, respectively.

A communication infrastructure encompassing the power plant model [2], real power plant data obtained from online databases (e.g., ERCOT [3]), and the control algorithms is established using the OSIsoft PI system at West Virginia University (WVU). The PI system centralizes the information received from the power plant model in MATLAB Simulink and the online power demand from the web, and send the control actions calculated by the MPC in MATLAB back to the power plant for implementation. To evaluate the proposed infrastructure, several days of power plant online data with cycling profiles are considered. The developed infrastructure enables power plant control using real-time data to improve decisions and operations of existing and future fossil-fuel-based power plants.

[1] He, X.; Lima, F. V. A modified SQP-based model predictive control algorithm: application to supercritical coal-fired power plant cycling. 2020. *Industrial & Engineering Chemistry Research*, *59*(35), 15671-15681.

[2] Agbleze, S.; Lima, F. V.; Indrawan, N.; Panday, R.; Pezzini, P.; Bonilla-Alvarado, H.; Bryden, K. M.; Tucker, D.; Shadle, L. J. Modeling and control of subcritical coal-fired power plant components for fault detection. *In Proceedings of the 2020 ASME ICONE 28-POWER*, August 2020.

[3] Electric reliability Council of Texas (ERCOT). (n.d.). Retrieved April 10, 2021, from http://www.ercot.com/

630337 CFD Based Study to Predict Rollover in an Industrial LNG Storage Tank

Abstract:

Natural Gas is transported in the liquefied form under cryogenic conditions, which makes LNG easier to transport and safer to store. LNG is stored in a highly insulated tank, despite which heat ingress from the environment is inevitable. Consequently, the lightest component (methane) evaporates. This results in a vapor stream known as Boil-off-gas (BOG) and increases the density of the remaining LNG. In an LNG terminal, when cargo is unloaded to a storage tank containing LNG of different densities, stratified layers with different liquid densities are formed. These layers further mix due to density equilibration and interface disruption under the effect of static pressure and cross the saturation temperature. This results in a large amount of BOG generation instantly, which leads to high tank pressure and a hazardous situation. This phenomenon of overall mixing and sudden generation of BOG is known as rollover. Over 24 rollover incidents have been reported over the last 60 years (GIIGNL, 2015), including those in La Spezia (Italy, 1971) and Partington (UK, 1993). In order to prevent the significant losses that arise during a rollover, it is important to develop a fundamental understanding of the boil-off-gas generation and the rollover phenomena, which is the focus of this paper.

The rollover phenomenon has been studied both theoretically as well as experimentally over the years. Germeles et al. (1975) proposed a double-diffusion LNG model by approximating methane & other components as a salt-sugar solution (thermohaline correlation) without considering boiling mechanism and turbulence during a rollover event. Heestand et al. (1983) improved this model by considering turbulence convection for heat and mass transfer between the stratified layers and rejected the salt-sugar solution model. Koyama et al. (2007), used a 3D CFD model to study rollover by considering light LNG filling only through the bottom of the tank without including the miscibility and the vapor mass removal from the liquid phase due to evaporation. Li et al. (2015) considered a 2D CFD model and induced stratification artificially to study rollover without considering the heat ingress. Hubert et al. (2019) improved the Li model by considering 3D CFD simulation with heat ingress through tank wall but did not include LNG filling mode of operation. Saleem et al. (2020) considered miscibility factors in the 2D CFD model and formulated stratification by different modes of tank filling. The major shortcoming of the model was to study the rollover under 2D tank without considering the effect of LNG unloading mode over rollover and BOG generation. These studies indicate that a 3D model of the tank that considers the effect of loading and unloading operations can represent the rollover in a more realistic manner. In the existing literature, rollover scenarios have been studied by inducing stratification artificially at the top and bottom of the tank, but the effect of unloading operation over rollover has not been studied. Since this is the most common occurrence in industry and the one of most concern to the LNG terminal operations personnel, we seek to fill these gaps in this work.

In this paper, we develop a CFD model to gain a fundamental understanding of the rollover phenomenon and the associated BOG generation. In this model, a 3D tank is used to store LNG at the receiving terminal. The tank is initialized to hold some initial level of LNG (inventory), while LNG of varying densities are allowed to fill either through the top or bottom of the tank to induce stratification. We consider heat ingress continuously through the walls of the tank, which increases the temperature of the stored LNG under the effect of local pressure (both tank top pressure and static pressure of LNG). This local pressure is used to calculate the bubble and dew point of the LNG mixture. Our model offers the flexibility to consider the LNG in the storage tank to be considered either as miscible or immiscible as necessary. In this paper, we report results from a case study to study rollover during LNG unloading operation. While this is a common operation in industry and the most susceptible to inducing rollover, it has not been previously studied in literature. Stratification is induced by filling a heavy density LNG from the bottom of the tank. Sensitivity studies on various important parameters tank size, inventory, rate of heat ingress, and density difference of LNG are also performed. Results from scaled-down studies (smaller size tank, shorter duration) and from industrial scales will both be presented. The operational benefits to the terminal from the accurate estimates of BOG generation rate and rollover time will also be discussed.

References:

1. GIIGNL Technical Study Group, 2015, Rollover in LNG Storage Tanks; Technical report.
2. Germeles, A.E., 1975. A model for LNG tank rollover 11.
3. Heestand, J., Shipman, C.W., Meader, J.W., 1983. A predictive model for rollover in stratified LNG tanks. AIChE J. 29, 199–207. https://doi.org/10.1002/aic.690290205
4. Koyama, K., 2007. CFD Simulation on LNG Storage Tank to Improve Safety and Reduce Cost, in: Koyamada, K., Tamura, S., Ono, O. (Eds.), Systems Modeling and Simulation. Springer Japan, Tokyo, pp. 39–43. https://doi.org/10.1007/978-4-431-49022-7_8
5. Li, Y., Li, Z., Wang, W., 2015. Simulating on rollover phenomenon in LNG storage tanks and determination of the rollover threshold. Journal of Loss Prevention in the Process Industries 37, 132–142. https://doi.org/10.1016/j.jlp.2015.07.007
6. Hubert, A., Dembele, S., Denissenko, P., Wen, J., 2019. Predicting Liquefied Natural Gas (LNG) rollovers using Computational Fluid Dynamics. Journal of Loss Prevention in the Process Industries 62, 103922. https://doi.org/10.1016/j.jlp.2019.103922
7. Saleem, A., Farooq, S., Karimi, I.A., Banerjee, R., 2020. CFD Analysis of Stratification and Rollover Phenomena in an Industrial-Scale LNG Storage Tank. Ind. Eng. Chem. Res. 59, 14126–14144. https://doi.org/10.1021/acs.iecr.0c02546

623091 Digital Twin of Renewable Energy-Linked Power-to-Gas (P2G) Systems: Model Building and Continuous Update for Hydrogen Producing Electrolysis

Abstract:

Solar energy and wind energy, which are widely used renewable energies, are highly volatile and intermittent, and long-term large-capacity electricity storage is required to maintain the stability of the power grid. Power-to-Gas (P2G) technology is used to generate and store gas from surplus electricity produced from renewable energy. Most P2G systems use electrolysis to produce hydrogen: the produced hydrogen is stored in a pressure tank and can be converted back to electricity through a fuel cell or hydrogen combustion engine when needed.

Hydrogen has a wide flammable range and low ignition energy compared to existing main energy sources such as gasoline or natural gas, so it has a high risk of fire and explosion, and is a material with high risk of emitting high combustion heat in case of explosion. Therefore, it is essential to evaluate the risk of hydrogen processing and storage equipment. The stability of the P2G system is an important factor in system operation along with safety. In particular, when the introduction rate of renewable energy increases, it is essential to operate a stable P2G system in order to process more excess energy.

This study aims to build a digital twin for a 500kW alkaline water electrolysis system to be built in the Saemangeum Renewable Energy National Demonstration Research Complex in Korea. We improve the digital twin model of the water electrolysis system using real-time sensing data enhanced by the real-time continuous model update. Though the current digital twin technology is focused on design and operation, we expand the applicability of digital twin by implementing functionalities for safe operation and safety management, including dynamic risk assessment, monitoring and fault detection.

In this study, based on the first principle-based models of alkaline water electrolysis cells, a digital twin of unit cells is developed, and a digital twin for the entire water electrolysis system consisting of a stack of these cells is developed. The core models consisting of the digital twin are as follows: 1) thermodynamic model for calculating Gibbs energy related to the reversible voltage, 2) electrochemical model for calculating the cell voltage according to the temperature, pressure and current, 3) electrolysis temperature calculation model, and 4) hydrogen production performance degradation model. The system/physics and failure/degradation models are composed of the above models, and operation data are combined to tune parameters of models in real-time, which establishes a digital twin of the alkaline water electrolysis system.

The developed digital twin is validated for the entire P2G system along with each of the cell units, alkaline water electrolysis system, and hydrogen compression and storage system. As a result, the developed digital twin is differentiated from the existing operation support system as a unified model for optimum control and scheduling, monitoring and fault detection, and dynamic risk assessment of the electrolysis system. When electricity production changes due to weather fluctuations, the development system enables quick start-up and system stabilization based on more accurate prediction and simulation, and dynamic risk management through prediction of functional degradation and damage of the entire system over time. It will be expected to enable real-time optimal operation and accident prevention of the alkaline water electrolysis system across the entire lifespan.

634115 Keynote Talk - Artificial Intelligence and the Modern Workforce

Abstract:

Owing to recent advances in Machine Learning and Artificial intelligence, companies across a wide range of industries are in a race to accelerate their digital transformation journey and become a data-driven enterprise where every decision and process becomes more effective and efficient, completed with higher levels of confidence, and yielding optimal results for company and customer alike. What does a successful employee of the future need to know? What will be the key characteristics of the most successful corporations? Why is it important to be a leader and not a laggard in this race? In this keynote, we will present real world examples from some of our customer engagements involving Industry 4.0 transformation. We'll highlight examples of large scale, industrial AI implementations and share key lessons learned as a preview of what's not only possible, but likely, in the future.

634552 On Demand Ceramic Open Cell For Intensified Reaction By High Precision Additive Manufacturing

Abstract:

By confining chemical reaction in a network of miniaturized structures in configurations measured in millimeters and embedded with micrometer-sized pores or channels, microarchitectured reactors provide more efficient mass and heat transfer because of their large specific surface areas, thereby resulting in a higher yield of reaction. A number of manufacturing technologies for chemically intensified reactors, including hot embossing, laser ablation, micromachining, and chemical etching, are currently commercially used. These techniques are generally constrained in design to two-dimensional planar channel networks, with more intricate designs leading to significant increases in cost of assembly, manufacturing complexity and production time. Additive manufacturing is emerging as a promising fabrication method because it allows precise control over the topology, geometry, and composition of the designed functional reactors.

In this invited talk, we introduce our research effort on additive manufacturing of cellular ceramic open cells with high-quality, complex architecture at critical dimension down to a few micrometers. This presentation highlights the background and engineering capabilities of high-resolution optical projection- based approach to additively produce complex polymer and ceramic parts on demand. We will also discuss our current research of ceramic materialization via chemical synthesis using printed polymeric cellular microstructures. Our proposed method overcomes the limitation of traditional ceramic manufacturing such as long lead time of tooling and high capital cost. By utilizing AM, catalytic converters far more geometrically complex than the traditional, extruded monolith structures can be manufactured. Therefore, additive manufacturing provides benefits outlined thus far for the catalytic converter. Using case studies of catalytic converter and reactors, we will discuss the outstanding challenges and fundamental issues that need to be addressed to bring high precision ceramics for intensified reaction at scale.

625089 Additive Manufacturing of Hierarchical Porous Structures for Controlled Gas Bubble Flows Using Cellular Fluidics

Abstract:

Porous architectures are widespread in natural and industrial systems and are useful for facilitating mass transport across gas/liquid interfaces. Advances in additive manufacturing (AM) have enabled the fabrication of highly complex porous structures with unprecedented feature size control over large build areas. Cellular fluidics is an AM-based approach for controlling the gas/liquid interface in a heterogeneous lattice on a unit-cell-by-unit-cell basis. In this presentation, we explore the transport of gas bubbles in cellular fluidic structures that are predominantly saturated with a liquid phase. Deterministic channels for gas flow are maintained by tuning the orientation and thickness of the beams comprising each unit cell. The walls of each channel are open, with flows being directed by the surface tension-derived pressure barrier between the gas and liquid phases. This allows for targeted delivery of gas flows with a high degree of spatial control while maintaining large interfacial areas for gases to dissolve into and extract out of the liquid phase. We demonstrate the applicability for these systems for the absorption of CO_2 and O_2. Cellular fluidic channels are also constructed in COMSOL, and the simulated pressure barrier across the gas/liquid interface is validated against what is observed experimentally.

625268 Cellular Fluidics: Tuning Multiphase Interfaces in 3D Using Architected Porous Media

Abstract:

Many processes in both nature and industry occur at the microscale and involve complex multiphase interfaces. Most microporous media, both natural and man-made, tend to be stochastic and therefore difficult to predict or control reliably. On the other hand, conventional microfluidic devices are often limited to enclosed channels and planar geometries, which hinders their usefulness in multiphase reaction or transport processes involving gas phases. We present a novel platform based on capillary fluid flow in ordered three-dimensional open-cell lattices. Using deterministic cell and lattice design, combined with additive manufacturing methods that provide access to length scales < 100 μm, we can fabricate complex 3D structures with tuned porosity and advanced functionalities. This approach enables selective placement and direction of liquid flow into predetermined continuous paths through the structure, as well as optimizing for the occurrence of gas-liquid, liquid-liquid, or gas-liquid-solid interfaces. We demonstrate the application of cellular fluidics for processes such as transpiration cooling, CO_2 capture, and selective patterning of catalytic materials for electrochemical CO_2 reduction. This work performed under the auspices of the U.S. Department of Energy by Lawrence Livermore National Laboratory under Contract DE-AC52-07NA27344 within the LDRD program 19-SI-005. LLNL-ABS-778327.

626896 CO_2 Capture Modeling with Intrastage Cooling

Abstract:

A new structured packing device has been designed recently at Oak Ridge National Laboratory that utilizes additive manufacturing. This device combines external gas-liquid contacting for mass transfer with internal coolant channels for heat transfer that results in process intensification for CO_2 capture with amine-based solvents. A process model has been developed to describe the intrastage cooling that takes place within this device. Experimental results with aqueous monoethanolamine (MEA) and with a low-aqueous solvent (LAS) are compared with this process model to demonstrate how CO_2 capture may be described within this novel structured packing device. The device achieves comparative separation efficiency as commercial structured packing while removing heat generated from reaction of CO_2 with amines. Further simulation work will be used to study process trade-offs between amine-based solvents to reach high CO_2 capture rates with intrastage cooling.

Thermal Energy Regulation with 3D Printed Polymer-Phase Change Material Composites

Peiran Wei,[a] Ciera E. Cipriani,[a] Emily B. Pentzer[a,b,*]

[a]Department of Materials Science and Engineering, Texas A&M University, College Station, TX, USA
[b]Department of Chemistry, Texas A&M University, College Station, TX, USA
*Correspondence: emilypentzer@tamu.edu

ABSTRACT

Phase change materials (PCMs) are an attractive option to passively control building heating and cooling. We present a facile method to produce and print PCM-filled inks by a direct ink writing (DIW) technique that leverages spherical PCM particles as viscosity modifiers in a matrix of photocurable resin. We successfully print inks with up to 63 wt% PCM, which have excellent thermal regulation capacity and nearly no leakage over 200 melting/solidifying cycles. A hollow house printed with PCM-filled ink maintained a 40% lower temperature than the external environment when heated. PCMs with different melting points can be simultaneously integrated into resin and printed without detriment to the structure or integrity. Our approach uses PCM particles as both viscosity modifiers for 3DP and passive thermal management to produce effective thermal buffers and is compatible with a wide range of photopolymer matrices and PCMs, without requiring prior microencapsulation of the PCMs.

INTRODUCTION

Increasing world population, economic growth, and industrialization result in higher global demand for energy. Currently, fossil fuel combustion is the most utilized source of energy for human needs.[1] In the U.S., 20% of energy is consumed for the active thermal control of buildings, in which indoor temperature variations are moderated by heating, ventilation, and air conditioning.[2] Unfortunately, the reliance on fossil fuels for these applications is met with resistance, both due to concerns of accessing such energy sources, as well as the negative environmental impacts of their combustion. As such, alternatives to active thermal control are a strategic target for sustainable global development, especially in the management of thermal energy. New technologies that reduce temperature fluctuations whilst maintaining desirable thermal comfort hold the potential to revolutionize the global energy landscape.

Thermal energy storage (TES) systems that utilize phase change materials (PCMs) for latent heat storage are attractive options for the passive control of building heating and cooling, as they can be efficient and environmentally benign.[3–8] PCMs absorb and release thermal energy through solid-solid, solid-liquid, or liquid-gas phase transformations within a well-defined temperature range.[3–5,9–11] These materials have proven useful in many applications, such as battery thermal management,[12] solar heating systems,[13–15] and cooling of electronics.[12,16,17] For example, Yin et al. produced paraffin/expanded graphite composite heat sinks for electronic components. These composites achieved apparent heat transfer coefficients 1.25 to 1.3 times higher than a traditional metal heat sink.[17] Alternatively, heat pipes, which typically consist of a hollow cylinder filled with a vaporizable liquid, are widely used as "thermal buffers" for lithium-ion batteries,[18,19] computer systems,[20] and battery systems of electric vehicles.[21] Selection of a PCM for a given application is based on cost, operating temperature, heat of fusion per unit weight, thermal conductivity, and cyclability.[4] Taking into account transformation temperature, energy storage density, and volume change, PCMs which undergo solid-liquid transformations are the most attractive choice for many systems, including thermal control of buildings[6,22] and batteries.[12,16]

To date, many solid-liquid PCMs have been developed and studied and can be classified as organic or inorganic. The most common organic PCMs are hydrocarbons, primarily paraffin waxes (C_nH_{2n+2}) and their fatty acid and ester derivatives.[3–5,7,9,10] On the other hand, inorganic PCMs, which include molten salts, metal alloys, and salt hydrates,[3–5,7–10] have recently received increased attention for their higher energy storage density and thermal conductivity, compared to organic PCMs. However, organic PCMs offer lower corrosivity and better compatibility with matrix materials, and they experience a lower degree of supercooling.[3–5,10] Ultimately, inorganic and organic PCMs offer complementary properties and should be selected based on the requirements of the intended application. Regardless of classification, all solid-liquid PCMs experience the limitations of loss of integrity and volume change upon melting. A further challenge in utilizing PCMs is the limited temperature range over which a single PCM performs its thermal regulation function.

A common approach to addressing the issues of leakage and volume change upon phase change is the microencapsulation of PCMs.[3,4,8,10,22–25] Several successful approaches have been demonstrated. For example, Advincula and coworkers reported the Pickering emulsion-templated encapsulation of stearic acid in a shell of graphene oxide (GO) nanosheets crosslinked by ethylenediamine. This core-shell structure prevented leakage of the molten PCM and improved its thermoregulation properties; further, these capsules were stable to multiple heating-cooling cycles.[23] In a similar vein, the Youngblood group encapsulated methyl laurate in a composite of hydrophobized cellulose nanocrystals and poly(urea-urethane), which formed rigid shells that inhibited leakage.[22] Such rigid microcapsules of PCMs can be integrated into building materials such as concrete, polymer binder, and gypsum to produce monolithic structures.[3,6,8,22–24] Cabeza et al. built a six-sided cubicle, with three of the walls containing 5 wt% of microencapsulated PCM in a concrete matrix. The PCM-loaded cubicle mitigated temperature fluctuations with a 1°C lower maximum temperature and 2°C higher minimum temperature than the control cubicle, and postponed the maximum temperature until two hours later than the control cubicle each day.[26] These findings demonstrate the benefits of the thermal buffer and thermal inertia effects of encapsulated PCMs when incorporated into building materials. However, for widespread integration and application, capsule formation must not have prohibitively complex manufacturing needs or costs, and the shell should have a limited impact on energy storage density.[23]

Building materials and infrastructure have benefitted from the recent progress in 3D printing (3DP) technologies. 3DP offers distinct opportunities to integrate active materials into monolithic structures, provided appropriate feedstock compositions can be realized. 3DP provides the ability to produce objects with complex geometries, such as decorative non-load bearing pieces, allowing for thermal energy storage materials to be incorporated into existing buildings. Among different 3DP techniques, direct ink writing (DIW) has become one of the most used strategies due to its low cost, ease of use, and the ability to tailor ink composition.[27] DIW inks must be shear-thinning and highly viscous to hold their shape after extrusion;[27] to date, ink compositions have included colloidal gels/suspensions,[28–30] polymers,[31] ceramics,[32,33] and nanoparticles.[34] As the viscosity of a colloidal suspension depends on the concentration of particle additives,[35,36] we hypothesized that particles of solid PCM can be used as viscosity modifiers to produce DIW inks. Such feedstocks would enable the rapid and scalable production of 3D printed functional monolithic structures with applications in thermal energy management of buildings without requiring a time-consuming encapsulation step beforehand.

Herein, we present a facile method to produce and print PCM-filled inks by DIW by leveraging spherical PCM beads as viscosity modifiers in a photocurable resin matrix. PCM beads were produced by emulsifying at elevated temperatures, then dispersed in commercially available acrylate resin, printed, and cured with ultraviolet light. As shown in **Figure 1A**, here PCM beads serve a twofold purpose of modifying ink rheology and imparting thermal energy management properties. The photopolymerization leads to elastic containment of the PCM without necessitating the use of a shell material. Based on this design concept, we successfully printed inks with up to 63 wt% PCM to give structures with excellent thermal regulation capacity and nearly no leakage during multiple heating and cooling cycles. Because ink formulation is independent of the chemistry of the PCM, multiple PCMs can be incorporated into a single ink; this allows for a wider operating temperature window and increases the thermal management capabilities of the structure (**Figure 1B**). We demonstrate that PCMs with different melting points (n-eicosane, paraffin wax, and n-hexatriacontane) can be simultaneously integrated into the resin and printed without detriment to integrity. Our method harnesses the advantages of DIW and eliminates issues inherent in current microencapsulation techniques – namely, fluid leakage upon volume change – to facilitate the incorporation of PCMs into building materials. The excellent adaptability of DIW makes it compatible with a wide variety of polymer matrix materials, offers the ability to tailor the loading level of different PCM particles to achieve desired thermal energy management performance, and gives control over the structure of the printed objects. This approach greatly simplifies the manufacturing process and decreases costs.

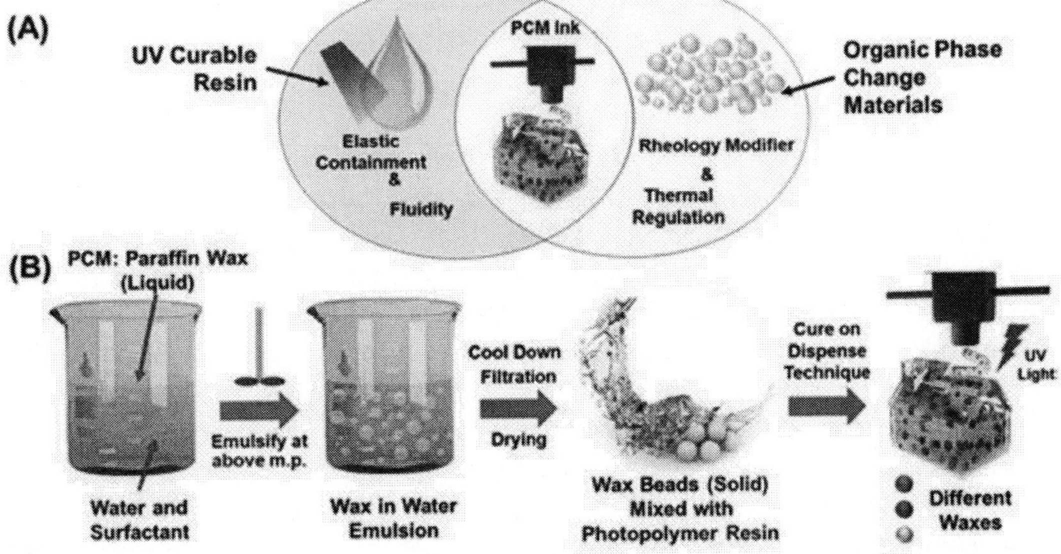

Figure 1. (A) Illustration of the design concept. The liquid resin offers an elastic containment to the phase-change material (PCM) and the PCM beads serve as rheology modifier. (B) Schematic illustrating the preparation and printing of the thixotropic ink composed of PCM beads and photopolymerizable resin.

RESULTS AND DISCUSSION

PCM Particles Impart Thixotropy to Photocurable Resin

Ideal inks for DIW should be thixotropic so that they are shear thinning and can be extruded, then quickly thicken to hold their shape. Typically, photocurable resins are Newtonian liquids, but nanofiller additives can impart non-Newtonian, thixotropic behavior.[35,37] We chose to use PCM microbeads as rheology modifiers. To this end, beads of paraffin wax (m.p. = 58-62°C) with an average diameter of 33 µm were prepared by emulsification at 80°C in water, in the presence of the surfactant Span® 20. **Figures S1-S2** show optical microscopy, nd scanning electron microscopy (SEM), and size distribution profile of the paraffin wax beads. These PCM beads were incorporated into a photocurable resin by hand mixing until a homogenous ink was produced (**Figure 2A**). Five photocurable resins were screened for ink formulation, and Formlabs Elastic resin was ultimately selected as it provided optimal rheological performance for printing and produced well-sealed PCM-containing structures that were stable to multiple heat-cooling cycles. Two additional PCM microbeads, n-eicosane (m.p. = 36-38°C) and n-hexatriacontane (m.p. = 74-76°C), were prepared and used as filler to achieve thermal regulation across different temperature ranges. Further, a non-PCM ink analog was prepared as a control by mixing cornstarch powder with the resin, since the resin itself is not viscous enough to be printed. For the data discussed below, the samples are named as form-filler-ratio in which: form is b = beads, i =ink, or c = cured ink; filler is P = paraffin, E = n-eicosane, H = n-hexatriacontane, or S = cornstarch; and the wt:wt filler:resin ratio of 0.7:1, 1.2:1, 1.7:1, and 2:1. For example, b-H refers to beads of n-hexatriacontane, i-P-1.7 refers to uncured paraffin ink with a weight ratio of 1.7:1 paraffin beads to resin, and c-R refers to the cured resin without filler.

The printability of inks with different PCM loadings was primarily evaluated by viscosity and yield stress. As displayed in **Figure 2B**, the as-received photocurable resin was a Newtonian liquid, as expected, with a viscosity that was independent of shear rate (black trace). Incorporating PCM beads into this resin imparted shear-thinning behavior, whereby the viscosity decreased as the shear rate increased. The loading of PCM beads was varied based on the weight ratio of wax to resin, from 0.7:1 to 2:1 (41 wt% to 67 wt% PCM, respectively; see **Table S1**). In general, increasing the filler concentration increased the viscosity, but had little effect on the magnitude of the shear-thinning behavior (*i.e.*, the slope). For example, at a shear rate of 4.09 s^{-1}, a viscosity of 7.4 $\times 10^4$ mPa·s was observed for i-P-0.7, and 6.5 $\times 10^5$ mPa·s was observed for i-P-2. This indicates that inks can be formulated to achieve optimal thixotropic behavior for DIW printing, since materials undergo shear during the extrusion process.

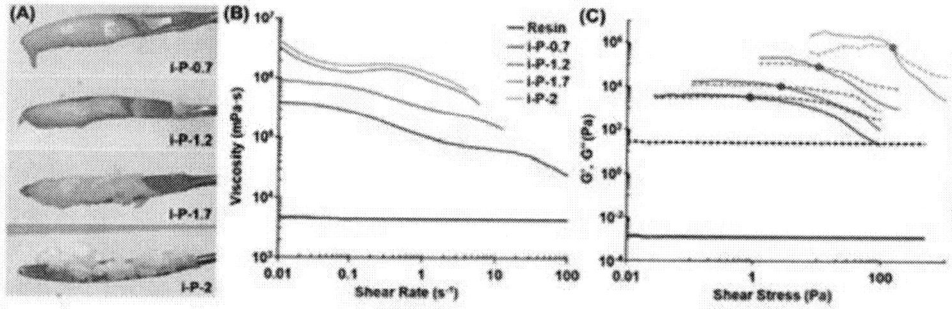

Figure 2. Rheological behavior of inks composed of elastic photocurable resin with different loadings of paraffin wax beads. (A) Digital images of inks with different loading of paraffin wax beads. (B) Apparent viscosity as a function of shear rate (data are truncated at the shear rates at which the ink aggregated and began to emerge from between the parallel plate). (C) Storage modulus and loss modulus as a function of shear stress.

As shown in **Figures 2C** and **S4**, increasing the loading of PCM beads increased the yield stress, which improved the ability of the ink to hold its shape after extrusion. For example, i-P-0.7 did not have a high enough yield stress (0.52 Pa) to hold its shape after extrusion; therefore, the ink dripped from the spatula. The samples i-P-1.2 and i-P-1.7 looked like pastes and retained their shapes on the spatula after inversion. However, further increasing the loading of PCM led to a granular ink that did not form a homogenous paste (*e.g.*, i-P-2). Thus, an upper limit to the loading of PCM exists for a composition to be printable. With this information, we chose i-P-1.2 and i-P-1.7 for printing, although similar, and intermediate, ratios should also be printable.

PCM Particles are Encapsulated by Photopolymer Resin in Printed Objects

Having established appropriate ink compositions for DIW printing, a Hyrel 3D Engine SR with a fixed layer height of 0.6 mm and an extrusion rate of 20 mL/h was used to print i-P-1.2 and i-P-1.7. After each layer was printed, UV light was used to cure the resin. Since the resin we used exhibits strong cyan fluorescence under 405 nm light and the PCM filler has no fluorescence, confocal microscopy was used to characterize the morphology of the cured PCM/resin composites. **Figure 3A** shows the printed and cured letters "T", "A", "M", and "U." **Figure 3B** shows the confocal microscopy image of c-P-1.7. The PCM particles appear as darker, spherical regions surrounded by a light gray color, which is the resin. As the loading of PCM particles increases, a higher concentration of darker regions is observed (see **Figure S5**). Breaking the c-P-1.7 sample and imaging the cross-section by SEM shows spheres and voids at the surface, as shown in **Figures 3C** and **S6**. These data support the presence of pockets of PCM surrounded by cured resin with the PCM particles homogeneously dispersed throughout the polymer matrix, suggesting that each particle is encapsulated within the monolithic structure.

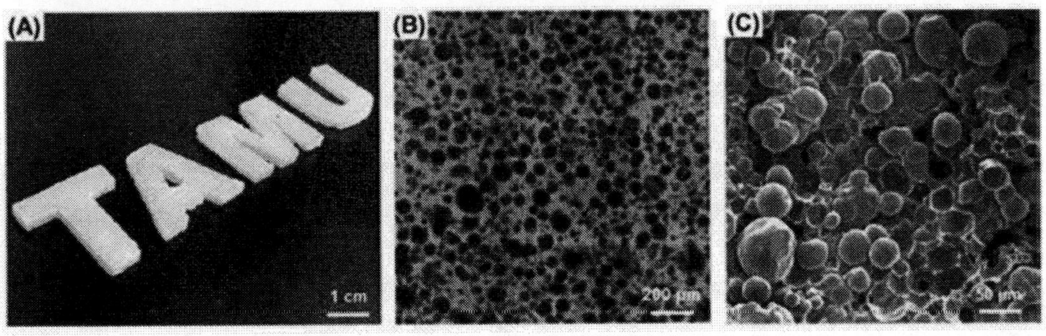

Figure 3. (A) Photograph of letters T, A, M, and U 3D printed and cured using PCM-containing inks. (B) Confocal microscopy image of c-P-1.7 (C) SEM image of the fractured cross-section of c-P-1.7.

Increasing PCM Loading Increases Elastic Modulus

We then evaluated the impact of ink composition on the mechanical properties. Dogbone-shaped tensile samples of each ink were cast according to ASTM Die C specifications and cured by UV light (**Figure S7**), then strained at a rate of

0.3 min^{-1} with a 1 kN load cell. A representative stress-strain curve of c-P-1.7 and a bar plot of the elastic moduli are shown in **Figure 4A** (details shown in **Table S2**). The elastic modulus and strain at break of c-R were 3.4 MPa and 43%, respectively, which indicates that the resin exhibits good elasticity. Compared to c-R, both the yield stress and modulus increased with increased loading of PCM beads (**Figure S8**). This behavior was expected and corresponds to the predictions of several models, including the modified Guth-Gold[38] and Takahashi[39] equations. The elastic modulus of c-P-2 (67 wt% PCM beads) was five times that of c-R. However, as the loading of PCM beads increased, the strain at break decreased accordingly (**Figure S9**). Thus, the PCM loading level dictates the mechanical properties of the resulting printed structure, and there is a tradeoff between elastic modulus and strain at break.

To evaluate the impact on thermal cycling on the mechanical properties of the samples, compression tests were conducted on as prepared c-R, c-S-1.7, and c-P-1.7 and after 100 and 200 thermal cycles when submersed in water. (**Figures S10** and **S11**). For the resin alone (c-R), the modulus was constant, whereas for c-S-1.7, a slight decrease in modulus (~10%) as the number of thermal cycles increased was observed. We attribute this to the hydrophobicity and thermal stability of the resin, and absorption of water by cornstarch filler, that latter of which causes swelling of the particles and poor adhesion between particle and matrix. In contrast, the c-P-1.7 samples increased in modulus as the number of thermal cycles was increased (~25%), which may be attributed to better filling of pockets in the cured resin by the PCM. All samples exhibited little change in compression behavior over the conditions tested Extensive thermal cycling appears to have little effect on the mechanical performance of c-R, c-S-1.7, and c-P-1.7.
Com

PCM Thermal Performance is Conserved within Printed Structures

The differential scanning calorimetry (DSC) thermal profiles of paraffin wax beads (b-P), printed and cured paraffin ink (c-P-1.7), and cured neat resin (c-R) are shown in **Figure 4B**; these data give insight into the sample composition as well as the impact of encapsulation on the wax beads. Both the pure wax beads and c-P-1.7 show a broad endothermic region from 45 to 65°C, with a peak at 55-57°C that corresponds to the paraffin melting temperature. This peak indicates that the wax beads within the cured resin have the same thermal behavior as the pure wax. Similar results were obtained for n-eicosane and n-hexatriacontane, whereby the same endothermic regions were observed for the PCM beads and inks (**Figure S12**). The DSC thermogram of pure resin is featureless, which corroborates the finding that the wax melting is responsible for any thermal energy management behavior of our materials. These data also support that pure domains of PCM exist within the cured resin, enabling the performance of the PCM to be conserved within the monolithic structure.

Increasing the loading of PCM in ink proportionally increases the amount of heat that can be stored and released by an object made from cured ink (**Figure S13**). Subjecting c-P-1.7 to 200 heating/cooling cycles revealed no change in the DSC profile (**Figure S14**), suggesting that the thermal management capability is stable and no morphological changes occur.

Paraffin Influences Thermal Expansion Only Above Melting Temperature

Thermal expansion is a common concern when incorporating PCMs into building materials. Paraffins are known to have a relatively large volume change when transitioning between the solid and liquid phases (~15%).[40] Therefore, we studied the linear thermal expansion of cured inks with different loadings of paraffin beads using a thermomechanical analyzer (TMA). **Figure 4C** shows the plot of dimension change of different inks versus time at a temperature ramp of 5°C/min after 5 minutes of equilibration at -25°C. Two distinct regions corresponding to different rates of dimensional change are identified. The resin had a constant linear coefficient of thermal expansion (CTE) over the range of temperatures tested. All cured paraffin-containing inks showed two regions of thermal expansion; the first region, between -20 and 25°C, is consistent across all samples and is attributed to the resin. For all inks containing PCM beads, a plateau in dimension change is observed from 45 to 65°C as the PCM melts. The second region of thermal expansion is between 72 and 78.5°C and is consistent across all PCM-containing inks. This region is not present in the resin. Thus, below the PCM melting point, the resin is the main contributor to CTE of cured inks, and above the PCM melting point, the paraffin melts and contributes to the material expansion. Therefore, an increased rate of dimension change and increased CTE is observed between 72 and 78.5°C. TMA data for each paraffin ink composition are available in **Figure S15**, and corresponding CTE values are listed in **Table S3**.

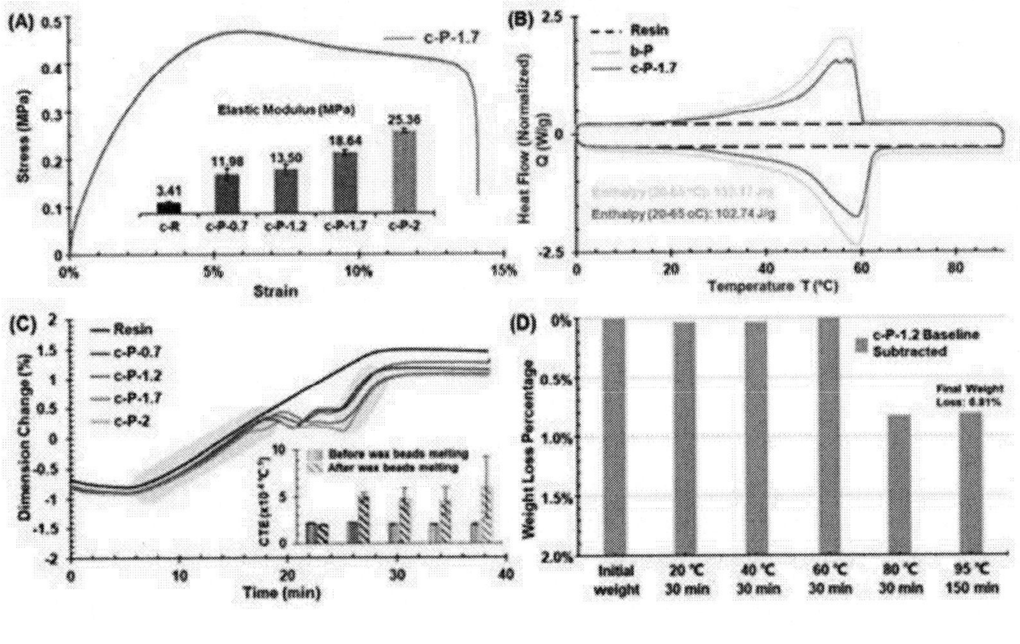

Figure 4. (A) Stress-strain curve of c-P-1.7 (inset compares the elastic modulus of inks with different loading of paraffin beads). (B) Representative DSC profiles of b-P, c-P-1.7, and c-R. (C) Dimension change of printed and cured structures as a function of temperature ramp (inset compares the CTE of c-R and inks c-P-0.7, c-P-1.2, c-P-1.7, and c-P-2 before and after the PCM melted). (D) Weight loss percentage of c-P-1.2 over the anti-osmosis (leakproofness) test in water at elevated temperatures.

Polymer-PCM Composites Exhibit Minimal PCM Leakage

The leakproofness of a PCM-containing structure, especially above the melting point of the PCM, is another important property for applying these materials in thermal energy management in buildings. Poor anti-osmosis performance may result in gradual permeation of liquid PCM out of the matrix (*i.e.*, a leaky system), resulting in a decrease in energy storage density over time. Therefore, we performed an anti-osmosis measurement to evaluate the leakproofness by submerging samples in water and increasing the temperature, as well as alternating between surfactant solutions and acidic and basic buffers. When immersed in water, the weight of both c-R and c-P-1.2 decreased slightly at room temperature (20°C), and this weight loss increased at elevated temperatures (**Figure S16**). To determine the weight of PCM lost, the weight loss of c-R was subtracted from the weight loss of c-P-1.2. As shown in **Figure 4D**, after baseline subtraction, the PCM-containing ink began losing weight in water only above 80°C, and < 1% weight loss was observed, even after heating in water at 95°C for more than 150 min. Similar experiments in the presence of surfactants and different pH (**Figure S17**), and reveal only ~0.8% weight loss after the first 100 cycles and ~1.8% after 200 cycles (**Figure S18**). Analysis of these samples by SEM imaging of the fractured surface indicates that, even after these experiments, the resin matrix maintained its porous structure, and the PCM particles are present (**Figure S19**). Thus, neither the presence of surfactant or changes in pH dramatically impact leakage of the PCM.

Different PCMs Provide Tailorable Thermal Performance

To evaluate the thermal energy storage capacity of our PCM-filled printed structures, the letters T, A, M, and U were 3D printed (**Figure 5**, for Texas A&M University). Each of the letters contained an ink with a different type of PCM bead, or a mixture thereof. Letter A was composed of c-E-1.2 (m.p. of n-eicosane is 36-38°C), letter M was composed of c-P-1.2 (m.p. of parrafin is 58-62°C), and letter U was composed of c-H-1.2 (m.p. of n-hexatriacontane is 74-76°C). Letter T was composed of c-EPH-1.2, which is an equal mixture of i-E-1.2, i-P-1.2, and i-H-1.2. A substrate for these letters was prepared from c-S-1.7 (i.e., the substrate was not expected to have thermal energy storage activity and provided a contrast to the PCM-containing letters). The design for the substrate and letters is shown in **Figure 20A**. **Figure S21** displays the printed letter "T" and a small printed substrate separately and in the assembled state.

The resolution of this printing process is shown by the optical image of cubic lattices printed with c-S-1.7 and c-P-1.7 in **Figure S22** and SEM images of printed c-P-1.7 in **Figure S23**. The extruded filaments are 0.85-0.9 mm with an interspace of 0.85-0.9 mm. In the high magnification images, protuberances can be found across the surface, indicating a high loading of PCM particles.

The thermoregulation performance of the printed PCM-filled structures was observed clearly through an infrared (IR) camera to visualize the different temperatures of the printed pieces. To evaluate these materials, the assembled printed samples were first cooled to -10°C, then transferred to a hotplate held at 100°C. As shown in **Figure 5**, as the temperature of the substrate increased to 57.3°C, the letter A showed a notably lower temperature (dark blue, **Figure 5iv**). As the temperature increased further, the n-eicosane beads in letter A fully melted; therefore, no difference in color between the letter and the substrate was observed in the thermal image. In this same image, the letter M is beginning this cycle as the substrate approaches the melting point of paraffin. From **Figure 5iv** to **5vii**, the c-S-1.2 substrate temperature increased up to 80.2°C while the letter M occupied a much lower temperature due to the absorption of thermal energy during the melting of the paraffin. Finally, the letter U began exhibiting thermal energy absorption in **Figure 5vi**, when the substrate approaches the melting point of n-hexatriacontane. The letters M and U maintain lower temperatures than the substrate even as the substrate reached 91.6°C. The letter T, which contained a mixture of beads of all 3 PCMs, also demonstrated thermal regulation ability by maintaining a lower temperature than the substrate across the entire temperature range tested. Thus, multiple PCMs can be combined within a single ink to increase the range of thermal regulation of printed structures.

When the heated letters and substrate were removed from the hotplate and allowed to return to room temperature, the letters exhibited the reverse behavior than that observed upon heating (**Figure S24**). The PCM-containing parts remained at higher temperatures as the c-S-1.2 substrate decreased in temperature. The letter U released heat first, shown in **Figure S24i-iii**, which corresponded to solidification of n-hexatriacontane. The letter M released heat next, shown in **Figure S24iii-vi**, corresponding to the freezing of paraffin. Finally, the letter A released heat during the freezing of n-eicosane, shown from **Figure S24v-ix**. The letter T, which contained all 3 PCMs, steadily released heat throughout the entire temperature range. Thus, the distribution of beads of PCM in structures enables the objects to possess excellent thermoregulation performance.

Increase temperature

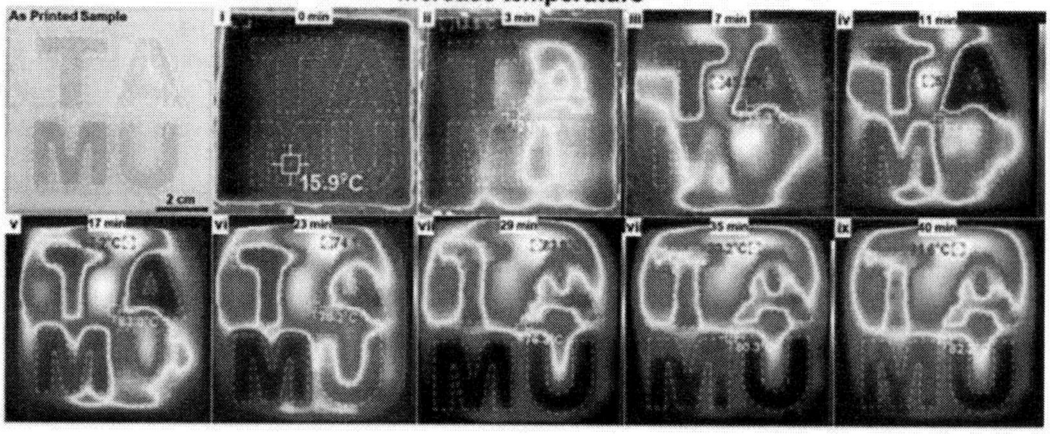

Figure 5. Representative infrared thermal images of the 3D printed and cured letters T, A, M, U inserted into the substrate composed of c-S-1.7. Letter A is c-E-1.2, letter M is c-P-1.2, letter U is c-H-1.2, and letter T is c-EPH-1.2. The sample was placed on a hot plate held at 100°C and allowed to increase in temperature.

PCM-Polymer Composites Have Superior Thermoregulation Capability to Conventional 3D Printing Materials

To further illustrate the impact of the incorporation of PCM beads into inks and printing of useful parts at scale, we printed and cured three small houses with commercial ABS, c-P-1.2, and c-S-1.2. Commercial ABS was used to demonstrate the performance of a typical thermoplastic used for fused filament fabrication 3D printing. The c-S-1.2 house allowed us to determine the thermoregulation capability of the resin matrix with no PCM, which was compared to that of the c-P-1.2 house. As **Figure S20B-D** shows, the printed houses were hollow and had approximately 10 mm thick walls. **Figure 6A** contains a schematic of the heating and temperature measurement setup: the houses were placed in an oven in which an electric fan circulated the hot air, and one thermocouple was inserted into each house while another thermocouple measured the chamber temperature. The chamber was heated to 80°C and held until the interior of every house reached approximately 80°C, at which point the heat was shut off, and the chamber door was opened to allow the houses to passively cool to room temperature. An image of this setup is provided in **Figure S25**. As shown in **Figure 6B**, during heating, the PCM-filled house exhibited a significant delay in temperature increase compared to the other two houses, which did not contain PCM. Alternatively, during cooling the PCM more slowly returns to ambient temperature, which correlates with the results of our IR camera experiment. **Figure 6C** shows that the PCM-containing house maintains a 40% lower temperature than the external environment, and 10-20% lower than the control houses during heating. Thus, the PCM-filled printed structures mitigate temperature fluctuations and can minimize the heating and cooling needed to maintain a constant temperature. The house made of c-P-1.2 also maintained its shape and structure, proving its stability at elevated temperatures.

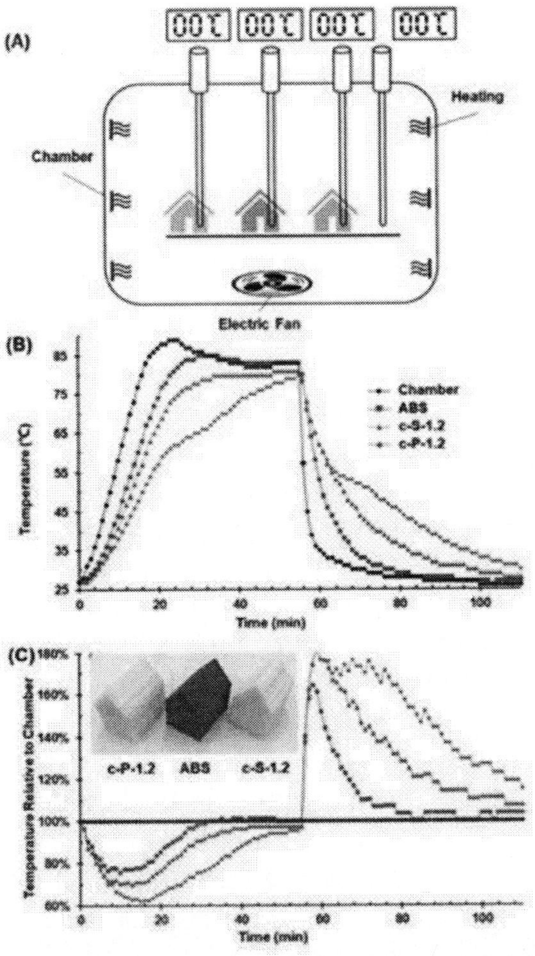

Figure 6. Thermal regulation capability of the houses printed with ABS, c-S-1.2 and c-P-1.2. (A) Schematic of how heating and temperature measurement was performed. (B) Variation with time of the temperature inside the houses during heating and cooling. (C) Temperature of inside of the houses relative to that of the chamber with time during heating and cooling (inset shows photograph of the houses).

CONCLUSIONS AND OUTLOOK

In summary, we have reported a facile method to produce and print PCM-filled inks by the direct ink writing (DIW) technique that leverages spherical PCM particles as viscosity modifiers in a matrix of photocurable resin. We evaluated the effect of ink composition on viscosity to optimize printability and performance in thermal energy management. Microscopy images of the cured structures revealed the homogeneous dispersion and full encapsulation of PCM beads within the resin matrix. A leakproofness test and DSC thermograms also demonstrated that the resin matrix effectively encapsulated the PCM throughout multiple solid-to-liquid phase change cycles, and that leakage of the molten PCM was negligible. The enthalpy of phase transformation was determined to be directly related to the loading of PCM beads. As ink formulation is not dependent on the identity of the PCM, multiple PCMs were incorporated into a single ink, allowing for a wider operating temperature window, and increased thermal management capabilities. These advances in ink formulation enabled us to 3D print hollow houses that served as effective thermal buffers across the melting temperature range of the PCM and with superior thermal buffer performance compared to structures without PCM. Our 3D printed houses mitigated temperature fluctuations with a 10% lower temperature during heating and 40% higher temperature during cooling than the houses without PCM. This method eliminates the need to microencapsulate PCMs before integration into a composite, enabling effective passive thermal management using readily available structural materials, and decreasing the manufacturing costs.

The excellent adaptability of DIW makes this approach compatible with a wide variety of polymer matrix materials, offers the ability to tailor the loading level of different PCM particles to achieve desired thermal energy management performance, and gives control over the geometry of the printed objects. This approach facilitates the production of PCM-containing structures with complex geometries, such as cooling fins in air conditioning units and replicas of architectural details for retrofitting existing buildings with thermal energy storage materials. The same design concept can be extended to other fields, such as the passive thermal management in spacecraft and electronics. Ongoing work focuses on diversifying the PCM filler used and optimizing resin selection to increase thermal conductivity and reduce the flammability of the composites.

References

(1) (2020). Annual Energy Outlook 2020 (Washington, DC).

(2) Pérez-Lombard, L., Ortiz, J., and Pout, C. (2008). A review on buildings energy consumption information. Energy Build. *40*, 394–398.

(3) Arshad, A., Jabbal, M., Yan, Y., and Darkwa, J. (2019). The micro-/nano-PCMs for thermal energy storage systems: A state of art review. Int. J. Energy Res. *43*, 5572–5620.

(4) Zhou, D., Zhao, C.Y., and Tian, Y. (2012). Review on thermal energy storage with phase change materials (PCMs) in building applications. Appl. Energy *92*, 593–605.

(5) Hawes, D.W., Feldman, D., and Banu, D. (1993). Latent heat storage in building materials. Energy Build. *20*, 77–86.

(6) Schossig, P., Henning, H.M., Gschwander, S., and Haussmann, T. (2005). Micro-encapsulated phase-change materials integrated into construction materials. Sol. Energy Mater. Sol. Cells *89*, 297–306.

(7) Kuznik, F., David, D., Johannes, K., and Roux, J.J. (2011). A review on phase change materials integrated in building walls. Renew. Sustain. Energy Rev. *15*, 379–391.

(8) Tyagi, V. V., Kaushik, S.C., Tyagi, S.K., and Akiyama, T. (2011). Development of phase change materials based microencapsulated technology for buildings: A review. Renew. Sustain. Energy Rev. *15*, 1373–1391.

(9) Akeiber, H., Nejat, P., Majid, M.Z.A., Wahid, M.A., Jomehzadeh, F., Zeynali Famileh, I., Calautit, J.K., Hughes, B.R., and Zaki, S.A. (2016). A review on phase change material (PCM) for sustainable passive cooling in building envelopes. Renew. Sustain. Energy Rev. *60*, 1470–1497.

(10) Shchukina, E.M., Graham, M., Zheng, Z., and Shchukin, D.G. (2018). Nanoencapsulation of phase change materials for advanced thermal energy storage systems. Chem. Soc. Rev. *47*, 4156–4175.

(11) Pielichowska, K., and Pielichowski, K. (2014). Phase change materials for thermal energy storage. Prog. Mater. Sci. *65*, 67–123.

(12) Zhao, R., Gu, J., and Liu, J. (2017). Optimization of a phase change material based internal cooling system for cylindrical Li-ion battery pack and a hybrid cooling design. Energy *135*, 811–822.

(13) Zhang, Y., Wang, S., Rao, Z., and Xie, J. (2011). Experiment on heat storage characteristic of microencapsulated phase change material slurry. Sol. Energy Mater. Sol. Cells *95*, 2726–2733.

(14) Garg, H.P., Mullick, S.C., and Bhargava, A.K. (1985). Solar Thermal Energy Storage (Dordrecht: D. Reidel Publishing Company).

(15) Guillot, S., Faik, A., Rakhmatullin, A., Lambert, J., Veron, E., Echegut, P., Bessada, C., Calvet, N., and Py, X. (2012). Corrosion effects between molten salts and thermal storage material for concentrated solar power plants. Appl. Energy *94*, 174–181.

(16) Kandasamy, R., Wang, X.Q., and Mujumdar, A.S. (2007). Application of phase change materials in thermal management of electronics. Appl. Therm. Eng. *27*, 2822–2832.
(17) Yin, H., Gao, X., Ding, J., Zhang, Z., and Fang, Y. (2010). Thermal management of electronic components with thermal adaptation composite material. Appl. Energy *87*, 3784–3791.
(18) Wu, M.S., Liu, K.H., Wang, Y.Y., and Wan, C.C. (2002). Heat dissipation design for lithium-ion batteries. J. Power Sources *109*, 160–166.
(19) Zhao, R., Gu, J., and Liu, J. (2015). An experimental study of heat pipe thermal management system with wet cooling method for lithium ion batteries. J. Power Sources *273*, 1089–1097.
(20) Singh, R., Akbarzadeh, A., Dixon, C., Mochizuki, M., and Riehl, R.R. (2007). Miniature loop heat pipe with flat evaporator for cooling computer CPU. IEEE Trans. Components Packag. Technol. *30*, 42–49.
(21) Rao, Z., Wang, S., Wu, M., Lin, Z., and Li, F. (2013). Experimental investigation on thermal management of electric vehicle battery with heat pipe. Energy Convers. Manag. *65*, 92–97.
(22) Yoo, Y., Martinez, C., and Youngblood, J.P. (2017). Synthesis and Characterization of Microencapsulated Phase Change Materials with Poly(urea-urethane) Shells Containing Cellulose Nanocrystals. ACS Appl. Mater. Interfaces *9*, 31763–31776.
(23) Advincula, P.A., De Leon, A.C., Rodier, B.J., Kwon, J., Advincula, R.C., and Pentzer, E.B. (2018). Accommodating volume change and imparting thermal conductivity by encapsulation of phase change materials in carbon nanoparticles. J. Mater. Chem. A *6*, 2461–2467.
(24) Memon, S.A. (2014). Phase change materials integrated in building walls: A state of the art review.
(25) Fang, Y., Kuang, S., Gao, X., and Zhang, Z. (2008). Preparation and characterization of novel nanoencapsulated phase change materials. Energy Convers. Manag. *49*, 3704–3707.
(26) Cabeza, L.F., Castellón, C., Nogués, M., Medrano, M., Leppers, R., and Zubillaga, O. (2007). Use of microencapsulated PCM in concrete walls for energy savings. Energy Build. *39*, 113–119.
(27) Lewis, J.A. (2006). Direct Ink Writing of 3D Functional Materials. Adv. Funct. Mater. *16*, 2193–2204.
(28) Xie, B., Parkhill, R.L., Warren, W.L., and Smay, J.E. (2006). Direct writing of three-dimensional polymer scaffolds using colloidal gels. Adv. Funct. Mater. *16*, 1685–1693.
(29) Smay, J.E., Cesarano, J., and Lewis, J.A. (2002). Colloidal inks for directed assembly of 3-D periodic structures. Langmuir *18*, 5429–5437.
(30) Sommer, M.R., Alison, L., Minas, C., Tervoort, E., Rühs, P.A., and Studart, A.R. (2017). 3D printing of concentrated emulsions into multiphase biocompatible soft materials. Soft Matter *13*, 1794–1803.

(31) Guo, Y., Liu, Y., Liu, J., Zhao, J., Zhang, H., and Zhang, Z. (2020). Shape Memory Epoxy Composites with High Mechanical Performance Manufactured by Multi-Material Direct Ink Writing. Compos. Part A Appl. Sci. Manuf. *135*, 105903.
(32) Minas, C., Carnelli, D., Tervoort, E., and Studart, A.R. (2016). 3D Printing of Emulsions and Foams into Hierarchical Porous Ceramics. Adv. Mater. *28*, 9993–9999.
(33) Lewis, J.A., Smay, J.E., Stuecker, J., and Cesarano, J. (2006). Direct ink writing of three-dimensional ceramic structures. J. Am. Ceram. Soc. *89*, 3599–3609.
(34) Weiß, M., Sälzler, P., Willenbacher, N., and Koos, E. (2020). 3D-Printed lightweight ceramics using capillary suspensions with incorporated nanoparticles. J. Eur. Ceram. Soc. *40*, 3140–3147.
(35) Wei, P., Leng, H., Chen, Q., Advincula, R.C., and Pentzer, E.B. (2019). Reprocessable 3D-Printed Conductive Elastomeric Composite Foams for Strain and Gas Sensing. ACS Appl. Polym. Mater. *1*, 885–892.
(36) Fang, Y., Yu, H., Wan, W., Gao, X., and Zhang, Z. (2013). Preparation and thermal performance of polystyrene/n-tetradecane composite nanoencapsulated cold energy storage phase change materials. Energy Convers. Manag. *76*, 430–436.
(37) Chen, Q., Cao, P., and Advincula, R.C. (2018). Mechanically Robust, Ultraelastic Hierarchical Foam with Tunable Properties via 3D Printing. Adv. Funct. Mater. *28*, 1800631.
(38) Guth, E. (1945). Theory of Filler Reinforcement. J. Appl. Phys. *16*, 20–25.
(39) Takahashi, K., Ikeda, M., Harakawa, K., Tanaka, K., and Sakai, T. (1978). Analysis of the Effect of Interfacial Slippage on the Elastic Moduli of a Particle-Filled Polymer. J Polym Sci Polym Phys Ed *16*, 415–425.
(40) Mann, A., Germann, T., Ruiter, M., and Groche, P. (2020). The challenge of upscaling paraffin wax actuators. Mater. Des. *190*.

624308 Elucidating Metabolic and Regulatory Mechanisms of Microbial Aromatic Utilization for Lignin Valorization

Abstract:

Lignin represents a renewable resource whose bioconversion could displace petroleum-based processes [1, 2]. Research has been conducted to develop a hybrid platform to generate value-added bioproducts from lignin breakdown products (LBPs) obtained by thermo-catalytic depolymerization of waste lignin [1-3]. Traditional model organisms are not well suited for converting LBPs that consist of various toxic aromatic compounds, but non-model organisms have been identified as ideal candidates. One such host is *Rhodococcus opacus*, which has demonstrated high native tolerance to LBPs and the ability to improve its tolerance and consumption through adaptive evolution. However, the key challenges in such lignin upgrading include our limited understanding of microbial utilization of toxic LBPs at gene levels and limited tools to engineer this organism. To understand and maximize its metabolic potential, we have employed multi-omics approaches, providing a systems-level understanding of the complex metabolism of the wild-type and evolved strains [4-8]. Additionally, we have developed a genetic toolbox for *R. opacus* engineering [9-12]. Despite recent advances in our understanding of its versatile metabolism and available genetic tools, studies of its gene functions at gene levels are still lagging.

In this presentation, we will discuss our recent efforts to facilitate functional studies of this non-model organism using our genetic toolbox and engineer this promising chassis for lignin valorization. Specifically, we have identified nine putative transcription factor (TF) genes in proximity to gene clusters for degradation of aromatic compounds and confirmed the function of these transcription factors using a *gfp* reporter gene and multiple deletion analyses. Interestingly, cross-interactions between different aromatic compounds' consumptions have been observed at the gene regulation levels, implying complex gene regulatory and metabolic networks of this organism. Additionally, we have developed and optimized a gene repression system based on CRISPR interference (CRISPRi) [11], and we will discuss our CRISPRi tool's utility by demonstrating the inducible accumulation of muconate from aromatics [13]. Altogether, this work represents a considerable achievement to understand and engineer this promising chassis for lignin conversion into value-added chemicals.

[1] Davis, K., Moon, T. S., Tailoring microbes to upgrade lignin. *Current opinion in chemical biology* 2020, *59*, 23-29.

[2] Anthony, W. E., Carr, R. R., DeLorenzo, D. M., Campbell, T. P., *et al.*, Development of Rhodococcus opacus as a chassis for lignin valorization and bioproduction of high-value compounds. *Biotechnology for biofuels* 2019, *12*, 192.

[3] Chatterjee, A., DeLorenzo, D. M., Carr, R., Moon, T. S., Bioconversion of renewable feedstocks by Rhodococcus opacus. *Current opinion in biotechnology* 2020, *64*, 10-16.

[4] Henson, W. R., Campbell, T., DeLorenzo, D. M., Gao, Y., *et al.*, Multi-omic elucidation of aromatic catabolism in adaptively evolved Rhodococcus opacus. *Metabolic engineering* 2018, *49*, 69-83.

[5] Yoneda, A., Henson, W. R., Goldner, N. K., Park, K., *et al.*, Comparative transcriptomics elucidates adaptive phenol tolerance and utilization in lipid-accumulating Rhodococcus opacus

PD630. *Nucleic Acids Res.* 2016, *44*, 2240–2254.

[6] Roell, G. W., Carr, R. R., Campbell, T., Shang, Z., *et al.*, A concerted systems biology analysis of phenol metabolism in Rhodococcus opacus PD630. *Metabolic engineering* 2019, *55*, 120-130.

[7] Henson, W. R., Hsu, F.-F., Dantas, G., Moon, T. S., Foston, M., Lipid metabolism of phenol-tolerant Rhodococcus opacus strains for lignin bioconversion. *Biotechnology for biofuels* 2018, *11*, 339.

[8] Hollinshead, W. D., Henson, W. R., Abernathy, M., Moon, T. S., Tang, Y. J., Rapid metabolic analysis of Rhodococcus opacus PD630 via parallel C-metabolite fingerprinting. *Biotechnology and bioengineering* 2016, *113*, 91-100.

[9] DeLorenzo, D. M., Henson, W. R., Moon, T. S., Development of Chemical and Metabolite Sensors for Rhodococcus opacus PD630. *ACS Synthetic Biology* 2017, *6*, 1973–1978.

[10] DeLorenzo, D. M., Moon, T. S., Construction of Genetic Logic Gates Based on the T7 RNA Polymerase Expression System in Rhodococcus opacus PD630. *ACS Synthetic Biology* 2019, *8*, 1921-1930.

[11] DeLorenzo, D. M., Rottinghaus, A. G., Henson, W. R., Moon, T. S., Molecular Toolkit for Gene Expression Control and Genome Modification in Rhodococcus opacus PD630. *ACS Synthetic Biology* 2018, *7*, 727-738.

[12] DeLorenzo, D. M., Moon, T. S., Selection of stable reference genes for RT-qPCR in Rhodococcus opacus PD630. *Scientific reports* 2018, *8*, 6019.

[13] DeLorenzo, D.M., Diao, J., Carr, R., Hu, Y., Moon, T.S., An Improved CRISPR Interference Tool to Engineer Rhodococcus opacus. ACS Synthetic Biology. doi.org/10.1021/acssynbio.0c00591

628399 Depolymerization of Corn Stover Lignin with Bulk Molybdenum Carbide Catalysts

Abstract:

Depolymerization of lignin into aromatic compounds in high yield for production of fuels and chemicals is vital to realize an economically competitive biorefinery. In this study, undoped and nickel-doped bulk molybdenum carbides were synthesized and evaluated in reductive depolymerization of lignin model compound and corn stover lignin in the presence of hydrogen. The highest lignin monomer yield of 37.3% was obtained with Ni-Mo_2C and ethanol/water solvent. Nickel doping promoted the depolymerization performance of Mo_2C, by increasing the number of metallic sites while decreasing that of acidic sites. This change in active site distribution mitigated CC coupling reactions and coking. Ethanol addition to the water solvent also significantly suppressed CC coupling. Mixing carbides with zeolites decreased monomer yields, presumably due to excessive repolymerization of reactive intermediates over the acidic zeolite. This work highlights the feasibility of developing effective lignin depolymerization catalysts by fine-tuning the properties of molybdenum carbide catalysts and solvent systems.

631798 Developing a Multi-Target Alkali Sterilization Strategy to Facilitate Lignin Dispersion and Promote Biological Lignin Valorization

Abstract:

Value-added utilization of lignin holds great promise for realizing sustainable and economic biorefineries. Biological lignin valorization represents an emerging green approach to upgrade lignin under mild operation conditions without intensive energy input. However, due to its complex three-dimensional aromatic structure, lignin generally exhibits poor water solubility and inhomogeneous distribution in the aqueous fermentation medium, limiting its microbial conversion efficiency significantly. The aim of this study is to develop a multi-target alkali sterilization strategy to effectively enhance lignin dispersion and fermentation performance. Furthermore, the colloidal particle size and molecular structure variations of lignin during the sterilization were thoroughly investigated to reveal the mechanisms of enhanced bioconversion performance. Results showed that a complete aseptic effect was realized when mixing lignin medium at an initial pH 12.7 for 24 h, demonstrating the effectiveness of the alkali sterilization strategy. Dynamic light scattering analysis implied that the volume of colloidal lignin particles decreased by 96.3% by alkali sterilization compared to the conventional thermal sterilization. Moreover, alkali sterilization reduced lignin weight-average molecular weight by 23.0% and narrowed the molar-mass dispersity by 23.8%. The lignin substrates with more uniform distribution and lower molecular weight benefited microbial cell growth and metabolism synchronously. Compared to conventional thermally sterilized lignin, alkali sterilized lignin medium promoted *Rhodococcus opacus* PD630 cell growth, lignin degradation, and lipids production, which were increased by 309%, 30.3%, and 48.3%, respectively. These results clearly implied that alkali sterilization dramatically improved the lignin bioconversion performance. Overall, this study presents a facile and effective sterilization strategy to overcome inhomogeneous lignin distribution in aqueous fermentation mediums, showing great potentials as a platform technique to promote biological lignin valorization.

631969 Demethylated Lignin As the Interface Enhancer in FDM 3D Printing for PA12 Composites

Abstract:

Fused deposition modeling (FDM) is a widely used 3D printing technique. One of the big challenges of FDM is the weak interface bonding among the layers which limits the mechanical properties of the printed items. In the current work, lignin, the most abundant aromatic biopolymer, was modified by converting its methoxy group to hydroxyl group, forming more phenolic OH groups. Quantitative ^{31}P NMR along with ^{1}H NMR technique was used to track the changes of phenolic OH and methoxy group, and results showed that the content of phenolic -OH group increased from 2.86 to 4.76 mmol/g while the -OMe group dropped from 15.66 to 5.00 mmol/g. The modified lignin (M-Lignin) was then applied in the FDM 3D printing with polyamide 12 (PA12). Due to the enhanced interaction between PA12 and M-Lignin, the melt viscosity increased significantly for the PA12/M-lignin blend during the filament extrusion. For the mechanical performance of the printed samples, the PA12/M-lignin composites with 10 wt% of the M-lignin had the highest tensile strength around 52.7 MPa (46.2 MPa for pure PA) and the Young's modulus of 1.98 GPa (1.69 GPa for pure PA12), while the elongation at break remained 46% (58% for pure PA). In addition, the M-lignin could also be potential to prevent the aging of PA12 in practical use.

632781 Biorenewable Chemical Production from Pyrolysate By Lactic Acid Bacteria

Abstract:

Lignocellulosic biomass is an abundant source of carbon, especially in agricultural areas like Iowa. Fast pyrolysis is one method that can be used to create fermentable sugars from cellulosic sources, but product of this process, bio-oil, is rich in (1) levoglucosan, an anhydrosugar form of glucose, and (2) phenolic and other inhibitory compounds. Here we report the engineering of *Lactococcus lactis* to utilize purified levoglucosan for the production of enantiomerically pure L-lactic acid and, by expressing a heterologous three-gene pathway, 1,2-propanediol. We also report media conditions that allow the production of these chemicals from diluted bio-oil and strategies to engineer tolerance to bio-oil and other "dirty" substrates. Finally, we report the development of CRISPR-assisted genome-scale engineering and library creation tools in this important industrial organism.

631978 Keynote Talk - Industry 4.0: Continuous Pharmaceutical Manufacturing Process

Abstract:

Currently, Industry 4.0 concepts are being applied to pharma industry to achieve Pharma 4.0 paradigm. Pharma 4.0 reduces the time and resources needed for continuous pharmaceutical manufacturing and also improves the product quality and production consistency. It has many advantages but also have bigger challenges on the applications of artificial intelligence (AI)/machine learning (ML), optimization, advanced process control and cyber-physical security side because of the different levels of complexities involved [1-2]. The predictive capabilities and the quality of the pharmaceutical products can be improved significantly via employing the artificial intelligence and the advanced model predictive control (MPC) system if an appropriate cyber-physical security defense is in place.

In this work, four components of industry 4.0 namely artificial intelligence (AI)/machine learning (ML), optimization, advanced control, and cyber-physical security, have been developed and implemented into the continuous pharmaceutical manufacturing process.

Four machine learning (ML) models have been trained to predict the response of continuous pharmaceutical manufacturing process and the performance of these ML models has been compared. The investigated ML methods are long short term memory (LSTM), 1D convolution neural network (CNN), random forest (RF), and artificial neural network (ANN). The best performing ML model is then implemented into the continuous pharmaceutical tablet manufacturing process for real time prediction.

A systematic framework including the methods and tools has been developed for dynamic optimization of the feeder refill strategies. The deviation of the outlet mass flow of the feeder from the targeted flow rate has been minimized to obtain the optimum value of the feeder refill parameters. The material properties also affect the refill strategy meaning that the feeder refill strategy need to be frequently optimized if there are any changes in the materials and plant. Therefore, the developed feeder model and dynamic optimization tool can save the time and recourses as well as can improve the product quality significantly.

An advanced model predictive control (MPC) system has been implemented in the continuous pharmaceutical manufacturing (CPM) pilot-plant. The CPP's and CQA's are controlled in real time. The critical control variables that have been controlled using model predictive control (MPC) system are drug concertation, powder level before tablet press, main and pre compression forces, tablet weight and hardness. A novel control strategy for powder level control in a chute placed in between blender and tablet press unit operation of continuous tablet manufacturing process has been developed, implemented and evaluated.

A systematic framework including the methods and tools have been also developed for proactive identification and mitigation of potential cyber-physical attack risk on continuous pharmaceutical manufacturing plant [2]. The cyber-physical security relevant software tools such as Snap 7, Wireshark, and Tripwire have been applied to CPM. A novel software tool named CPS (Cyber-Physical Security) has been developed for cyber-physical security of the continuous pharmaceutical manufacturing. The integrated commercially available and developed (in house) cyber-physical security tools have added an extra layer of security of our continuous pharmaceutical manufacturing pilot-plant for any unexpected attacks.

All the relevant data generated during continuous manufacturing has been systematically collected, stored and organized in a data hub (OSI PI) and cloud system as per industry 4.0 standard.

References

[1]. Bhaskar, A., Barros, F. N., Singh, R. (2017). Development and implementation of an advanced model predictive control system into continuous pharmaceutical tablet compaction process. International Journal of Pharmaceutics, 534 (1-2), 159-178.

[2]. Singh, R. (2020). The cyber-physical security of pharmaceutical manufacturing processes. Pharma., Issue 38, 53-57.

623634 Adaptive Strategies for Updating Unit Operation Models and in-Line Monitoring of Blend Uniformity in Continuous Pharmaceutical Manufacturing Process

Abstract:

Guided by U.S. Food and Drug Administration (FDA)'s Quality-by-Design (QbD) initiative[1,2] and market incentives[3], the pharmaceutical industry has heavily invested in the transition of tablet production from a traditional batch operating mode to continuous pharmaceutical manufacturing (CPM) over the past decade. The benefits of employing CPM, including but not limited to the smaller footprint, reduced human impacts, increased controllability, and shortened release time, have been clearly identified[4]. To better understand this emerging process and perform quality assessments, different mathematical models for unit operations and in-line process analytical technology (PAT) tools have been developed[4,5]. On the unit operation level, depending on the availability of process knowledge and the extent of data incorporation, these models can be classified as: (1) mechanistic models that strongly rely on process understanding; (2) data-driven models which depend only on the process data; and (3) hybrid models that integrate the first two categories[6,7]. In terms of the in-line PAT tools which monitor product quality, the performance relies heavily on data-driven global calibration models, such as partial least squares (PLS) regression[8].

The development and implementation of data-driven and hybrid models have a well-known challenge related to model maintenance. Since these models are typically trained based on historical data, they can only reflect the system for the range of process conditions, material properties, and environmental variables present at the time of model building[9]. However, shifts in the CPM process can happen over time due to changes such as environmental conditions, equipment wear and tear, lot-specific material properties, operating conditions, etc. These changes can happen suddenly, gradually, incrementally, or in a pattern[10], impacting performance of unit operations and ultimately, the critical quality attributes (CQAs) of a product. It is noted that these changes are usually not captured in typical training datasets, leading to deteriorating model prediction accuracy[10]. Therefore, appropriate model maintenance strategies need to be established. For unit operation models that involve data-driven components, model maintenance can ensure accurate representation of the manufacturing process and allow dynamic system analyses. For PAT-based chemometric models, maintenance is important for the calibration models to be robust and transferable[11,12], but traditional maintenance strategies are generally complex, costly, and time-consuming[8,11,13]. For example, depending on the diagnostics of a reduced model performance, more calibration data may need to be added, requiring additional experiments to be performed, consuming time and financial resources.

In this work, adaptive modeling strategies are developed and assessed, to serve as a model maintenance tool to cope with non-linear time-varying changes in process characteristics and/or operating conditions. The developed hybrid unit operation models for the continuous direct compaction (CDC) line[4,14] are used as baseline models. Adaptive windowing[15] and ensembled adaptive windowing[15] are selected as adaptive algorithms to be embedded into the data-driven components of the base cases. In addition to using inputs at specific time points for prediction, these algorithms incorporate change detection mechanisms, which monitor the process data streams to identify any process changes. Given such information, the algorithms can then update the data-driven components of the models. The prediction performances of the resulting adaptive models are compared against the base models in cases involving changes in process conditions, material properties, and design parameters. In terms of the PAT-based chemometric models, performance of locally weighted PLS (LW-PLS)[16], moving window[15], and recursive PLS (RPLS)[17] are evaluated in comparison to regular PLS for in-line monitoring of low dose blends in the feed frame of a tablet press using near infra-red (NIR) spectroscopy. Performance assessment involves a cross-prediction between datasets comprising of NIR spectra collected at different feed frame speeds with physical baseline differences in the NIR spectra. Cross-prediction can help to evaluate the effectiveness of adaptive methods to prevent the anticipated deterioration of prediction accuracy. Case studies involving time-varying process condition changes are also developed for comparison. The results of adaptive unit operation and chemometric models demonstrate a more accurate prediction performance in systems with changes in operating conditions. Using these adaptive modeling strategies, the models can be updated in real-time or periodically to enhance the predictability of CQAs throughout the process development and operation phases[18].

References

(1) Yu LX, Akseli I, Allen B, et al. Advancing Product Quality: a Summary of the Second FDA/PQRI Conference. *AAPS J.* 2016;18(2):528-543.

(2) Yu LX, Kopcha M. The future of pharmaceutical quality and the path to get there. *Int J Pharm.* 2017;528(1-2):354-359.

(3) Boukouvala F, Niotis V, Ramachandran R, Muzzio FJ, Ierapetritou MG. An integrated approach for dynamic flowsheet modeling and sensitivity analysis of a continuous tablet manufacturing process. *Comput Chem Eng.* 2012;42:30-47.

(4) Wang Z, Escotet-Espinoza MS, Ierapetritou M. Process analysis and optimization of continuous pharmaceutical manufacturing using flowsheet models. *Comput Chem Eng.* 2017;107:77-91.

(5) Ierapetritou M, Muzzio F, Reklaitis G. Perspectives on the continuous manufacturing of powder-based pharmaceutical processes. *AIChE Journal.* 2016;62(6):1846-1862.

(6) Zendehboudi S, Rezaei N, Lohi A. Applications of hybrid models in chemical, petroleum, and energy systems: A systematic review. *Applied Energy.* 2018;228:2539-2566.

(7) von Stosch M, Oliveira R, Peres J, Feyo de Azevedo S. Hybrid semi-parametric modeling in process systems engineering: Past, present and future. *Comput Chem Eng.* 2014;60:86-101.

(8) Wise BM, Roginski RT. A Calibration Model Maintenance Roadmap. *IFAC-PapersOnLine.* 2015;48(8):260-265.

(9) Webb GI, Hyde R, Cao H, Nguyen HL, Petitjean F. Characterizing concept drift. *Data Mining and Knowledge Discovery.* 2016;30(4):964-994.

(10) Gama J, Žliobaitė I, Bifet A, Pechenizkiy M, Bouchachia A. A survey on concept drift adaptation. *ACM Computing Surveys.* 2014;46(4):1-37.

(11) Zeaiter M, Roger JM, Bellon-Maurel V, Rutledge DN. Robustness of models developed by multivariate calibration. Part I. *TrAC Trends in Analytical Chemistry.* 2004;23(2):157-170.

(12) Besseling R, Damen M, Tran T, et al. An efficient, maintenance free and approved method for spectroscopic control and monitoring of blend uniformity: The moving F-test. *J Pharm Biomed Anal.* 2015;114:471-481.

(13) Zhong L, Gao L, Li L, Zang H. Trends-process analytical technology in solid oral dosage manufacturing. *Eur J Pharm Biopharm.* 2020;153:187-199.

(14) Chen Y, Ierapetritou M. A framework of hybrid model development with identification of plant-model mismatch. *AIChE Journal.* 2020.

(15) Kadlec P, Grbić R, Gabrys B. Review of adaptation mechanisms for data-driven soft sensors. *Comput Chem Eng.* 2011;35(1):1-24.

(16) Hazama K, Kano M. Covariance-based locally weighted partial least squares for high-performance adaptive modeling. *Chemometrics and Intelligent Laboratory Systems.* 2015;146:55-62.

(17) Joe Qin S. Recursive PLS algorithms for adaptive data modeling. *Comput Chem Eng.* 1998;22(4-5):503-514.

(18) Gama J, Medas P, Castillo G, Rodrigues P. Learning with Drift Detection. 2004; Berlin, Heidelberg.

626250 Data-Driven Approaches Towards Equipment Health-Classification and Predictive Monitoring in Drug Product Manufacturing

Abstract:

Digitalization is the central pillar towards the introduction of Industry 4.0 in pharmaceutical manufacturing. However, sophisticated data-driven approaches are yet to be explored to fully take advantage of the abundance of recorded process data available. Drug manufacturing is a highly regulated field. Thus, regulators require manufacturers to store process data over many years for backtracking purposes. This work explores the potential to use such data for predictive maintenance purposes.

This work presents a data-driven approach for equipment monitoring and the implementation of equipment classification and predictive maintenance in an aseptic filling line in drug product manufacturing. The goal is to leverage stored short-term process monitoring data for a secondary application in long-term equipment monitoring. Multiple years' worth of industrial process data are used, which include process monitoring data and logbook information regarding each process run. Process monitoring data is mainly acquired through Supervisory Control and Data Acquisition (SCADA). Monitored variables include pressure, temperature, or process step duration. Furthermore, logbook information records details regarding equipment and process failures as well as maintenance data including measures taken and maintenance intervals.

In a first step, process data was pre-processed and different classification algorithms were applied. Thereby, equipment failures were shown to be clearly linked with process monitoring data through, either, observable shifts in principal curves or peaks representing events. Short-term noise in the process data through variation caused by manual operation and batch-wise production were successfully minimized and long-term based effects due to equipment deterioration were identified. An analysis of results obtained from different classification algorithms for equipment monitoring is presented.

In a next step, different machine learning algorithms are applied to the process data and compared in their ability to predict equipment deterioration and failure. Additionally, optimized maintenance strategies for equipment are formulated. Finally, a decision-support framework is being developed, where expert knowledge will be included for defining counter measures taken in the case of alarming changes in equipment status.

630095 An Equation-Oriented Model of a Trickle-Bed Reactor for Hydrodesulfurization Process Analysis and Digital Twin Applications

Abstract:

The Fourth Industrial Revolution is marked by the convergence of digital, physical, and biological technologies, bringing relevant tools to allow deep process analysis and optimizations, thus enabling more assertive decision-making. One such tool directly linked to rigorous modeling and simulation is the process digital twin. When a digital twin is connected to powerful data sources (i.e., rigorous thermodynamic basis, industrial data sets, and detailed phenomenological models), it can be used for process design [1,2]. In this sense, the industrial sector, particularly the oil and gas industry, can benefit of digital twin approaches to improve process efficiency. The Hydrodesulfurization (HDS) process is considered a well-known and established technology used in refineries to remove sulfur compounds, such as sulfides and disulfides, from petroleum fractions. HDS consists of a catalytic process capable of treating a contaminated oil charge stream through a hydrogenation reaction under controlled operating conditions that are highly dependent on the oil physicochemical characteristics [3]. However, the definition of such operating conditions, as well as the evaluations related to reactant characteristics and catalysts need special attention.

In this work, the improvement of a rigorous HDS process model, based on a trickle-bed reactor block [4] in AVEVA Process Simulation is presented. The previously validated phenomenological model [5,6] has been connected to the AVEVA thermodynamics database for comprehensive process analysis. The resulting simulation enables the description of the sulfur removal and the formation of H_2S process throughout the reactor, in addition to satisfactorily predicting the effects on the conversion of the organic sulfur compounds through changes in nominal trickle-bed reactor conditions. The obtained relationship employing this model between the gas/oil ratio and the partial pressure of hydrogen sulfide presented a realistic behavior, enabling improvements on sulfur removal. Operating conditions, such as temperature and pressure, present consistent performance associated with effectiveness of desulfurization. The concentrations of the compounds in liquid phase and on the catalyst surface are analyzed throughout the reactor confirming experimental and industrial results [5,6].

References

[1] Macchi, M., Roda, I., Negri, E., & Fumagalli, L. (2018). Exploring the Role of Digital Twin for Asset Lifecycle Management. IFAC-PapersOnLine, 51(11), 790–795.

[2] Rosen, R.; Von Wichert, G.; Lo, G., & Bettenhausem, K. D. (2015). About The Importance of Autonomy and Digital Twins for the Future of Manufacturing. In IFAC-PapersOnLine, 48:567–72. Elsevier Ltd.

[3] Mendez, F.J.; Franco-Lopez, O.E.;. Bokhimi, X.; Solis-Casados, D. A.; Escobar-Alarcon, L., & Klimova, T. E. (2017). Dibenzothiophene hydrodesulfurization with NiMo and CoMo catalysts supported on niobium-modified MCM-41, Appl. Catal. B Environ. 219, pp 479–491.

[4] Barros, A.; Bishop, B..; Tavernard, A.; Lima, F. V., & Bispo, H. (2020). Development of an Equation Oriented Model Applied to a Hydrodesulfurization (HDS) Process. AIChE Annual Meeting.

[5] Neto, A. T. P., Fernandes, T. C. R. L., da Silva Junior, H. B., de Araújo, A. C. B., & Alves, J. J. N. (2020). Three-phase trickle-bed reactor model for industrial hydrotreating processes: CFD and experimental verification. Fuel Processing Technology, 208.

[6] Korsten, H., & Hoffmann, U. (1996). Three phase reactor model for hydrotreating in pilot trickle bed reactor, AIChE Journal, 42, pp 1350-1360.

634116 Keynote Talk - Digital Transformation, Data Science, and Industrial Intelligence

Abstract:

Data analysis has played an important role in process monitoring and operations for over three decades, with early methods focusing on multivariate statistics and artificial neural networks. Over the last ten years, along with the rise of big data, industrial internet of things (IIoT), statistical machine learning, and deep neural networks, digitalization and data analytics have taken place in nearly all industry sectors with an unprecedented pace. The McKinsey report in 2011 envisioned that analyzing large data sets would become a key basis of competition, underpinning new waves of productivity growth and innovations. Recent surveys by McKinsey revealed that the adoption of machine learning varied significantly from sector to sector.

In this keynote we offer a few perspectives on the development of data analytics and machine learning in process systems engineering and how we can maneuver their healthy development to harness the full potential of the digital revolution. Digitalization and the IIoT adoption will provide the process industries with massive amount of data from process operations to process maintenance. The power of data science and analytics from statistical learning methods to deep neural networks will need to be scrutinized and tailored for process applications. In driving toward industrial intelligence, we attempt to address the following big-picture questions in this talk.

i) What is the role of data science and analytics in transforming data into information and knowledge?

ii) Will human be out of the loop and replaced by artificial intelligence or autonomous systems?

iii) Will industry need deep learning or explainable machine learning?

iv) Will first principles be replaced with data-driven deep learning?

634131 Keynote Talk - Decision Making in Digitalized Value Chain Optimization

Abstract:

With digitalization being such a hot topic in control and real-time optimization, this talk focuses more narrowly on tactical and strategic decision making which typically has more humans in the loop. After briefly introducing ExxonMobil's Downstream value chain, I will highlight two broad themes: Modeling competition between players and tractable optimization under uncertainty. In the first theme, I will introduce problems in which multiple players vie with one another to maximize individual profit in a non-cooperative competitive market. This competitive setting is interesting and worthy of study because the majority of prevailing process systems engineering models largely overlook non-cooperative strategies that exist in real-world markets. In the second theme, I will discuss the decision rule approach to optimization under uncertainty and how we have applied it to several PSE problems. I will conclude with some future research topics that I hope are of interest to the broad community.

629856 3D Stereolithography (SLA) Printing-Based Micro-Fabrication Using Custom Polymer Resin Chemistry for Rapid Prototyping of Microfluidic Chips and Component

Abstract:

We report the use of an off the shelf and inexpensive desktop masked stereolithography (MSLA) based 3D printer to (i) fabricate complex structures with sub 100 micron-sized features with 3D printed molds for conventional PDMS casting, (ii) fabricate unibody chips with imbedded channels, and (iii) Engineer a finger-actuated displacement pump that can transport and meter precise quantities of fluids into a microfluidic chip. Parameters such as print direction, support structure, and light exposure time were found to be important factors that determine the optimal conditions to achieve the smallest possible print resolution consistently. We also present a custom UV curable resin that enables us to fabricate transparent prints with tunable compressibility and porosity. We demonstrate the fabrication of multi-layered microfluidic channels, droplet generators, and DNA amplification chips. The convenience and swiftness of our proposed technique enables one to go from an “idea stage” to a “fabricated microfluidic chip” within ~hours as opposed to traditional clean room-based microfabrication (~days), which could drastically lower the barrier to entry and greatly accelerate lab-on-a-chip research.

631822 Mapping Process-Induced Chain Orientation in 3D Printed Parts

Abstract:

Material extrusion additive manufacturing is characterized by a complex processing history that includes high deformation rates and rapid cooling. This process history is known to trap chains in the extruded polymer melt into nonequilibrium conformations, and the resulting molecular-scale anisotropy correlates well with variations in part properties like the weld strength. Residual chain orientation is commonly measured via optical birefringence through the transmitted part. Quantitative birefringence measurements can be time-consuming, and refraction at the polymer-air interface can obscure measurements near the weld, where orientation is expected to be significant. Here, we show that a combination of polarized optical imaging and optical microscopy can be used to map chain orientation in printed parts. Polarized optical imaging is performed on printed polylactide parts in refractive index matched fluids using a pixelated polarizer array bonded to the image sensor of a camera. The use of a refractive index matched fluid minimizes refraction to capture both bulk and near-weld information. The resulting two-dimensional image quantifies the orientation direction and the retardance, defined as the product of the birefringence and the local sample thickness. When combined with thickness measurements from optical microscopy, the polarized images provide a map of birefringence, which is proportional to the chain orientation. This orientation is highly sensitive to processing conditions like the print speed and nozzle temperature, which can be directly related back to two critical timescale comparisons: 1) the Weissenberg number comparing flow timescales in the nozzle to the reptation time in the polymer melt, and 2) the comparison between the reptation time necessary to relax chain orientation and the time required for the polymer to cool to the glass transition temperature. At high Weissenberg numbers and fast cooling conditions relative to chain relaxation, chains highly oriented in the flow direction should remain trapped in the glassy state, which is directly observed in these experiments. These results show the importance of optical polarization imaging for part characterization and potential quality control.

630951 Two-Wavelength Volumetric 3D Printing for Rapid Fabrication of Multi-Level Microfluidic Networks

Abstract:

Conventional microfluidic fabrication faces multiple challenges which prevent scalable manufacturing. Microfluidic fabrication is often done using lithography, which results in long fabrication time and high costs, as well as confining the device to two-dimensional geometries. Recently, 3D printing has emerged as an effective solution to scale up microfluidic fabrication. However, 3D printing often involves discrete layer-by-layer steps, which result in slow print times and still limits the range of printable geometries. In this work, we demonstrate a novel volumetric 3D printing method for rapid fabrication of microfluidic networks. Our approach uses two wavelengths of light to generate simultaneous inhibition and initiation reactions in free-radical polymerized resins. (Meth)acrylic resins are formulated with a unique system of photoinitiator and photoinhibitor species such that irradiation by near UV light (365 nm) inhibits the polymerization reaction and irradiation of blue light (460 nm) initiates polymerization. Polymerization and inhibition models are calculated as a function of light dose and used to predict the depth of polymerization and inhibition. Superimposing UV light with patterned blue light enables spatial control of polymerization in three dimensions without a moving build platform.

Using this system, we fabricated microfluidic networks in less than 15 minutes. Initially, we used polymerization and inhibition models to calculate the necessary parameters, such as light dose, to print at various levels in the z-direction. Next, to print the networks, resin was filled between two glass slides to create a resin vat, and a sequence of UV and patterned blue exposures were projected into the resin vat. Finally, the print was post-treated by flushing uncured resin from the channels and irradiating the part with visible light, and the glass slides were removed to create a monolithic microfluidic network. The quality of the printed channels was investigated using optical profilometry. Our initial data suggests that we can control polymerization as far as 700 μm into the resin in the vertical z-direction. The profilometry results agree with our predictive models and illustrate the necessity of the models to print accurate geometries. Different geometries including multi-level serpentine mixing channels and vertically overlapping channels were printed. We successfully show that our system is well suited for rapid, low-cost, microfluidic fabrication in three dimensions. This technology represents a significant advance in both resolution and functionality over traditional lithography-based fabrication techniques which could dramatically lower the barrier-to-entry to explore microfluidic systems, thereby enabling accelerated microfluidic innovation and application.

Figure Caption: Schematic of the two-wavelength volumetric 3D printer.

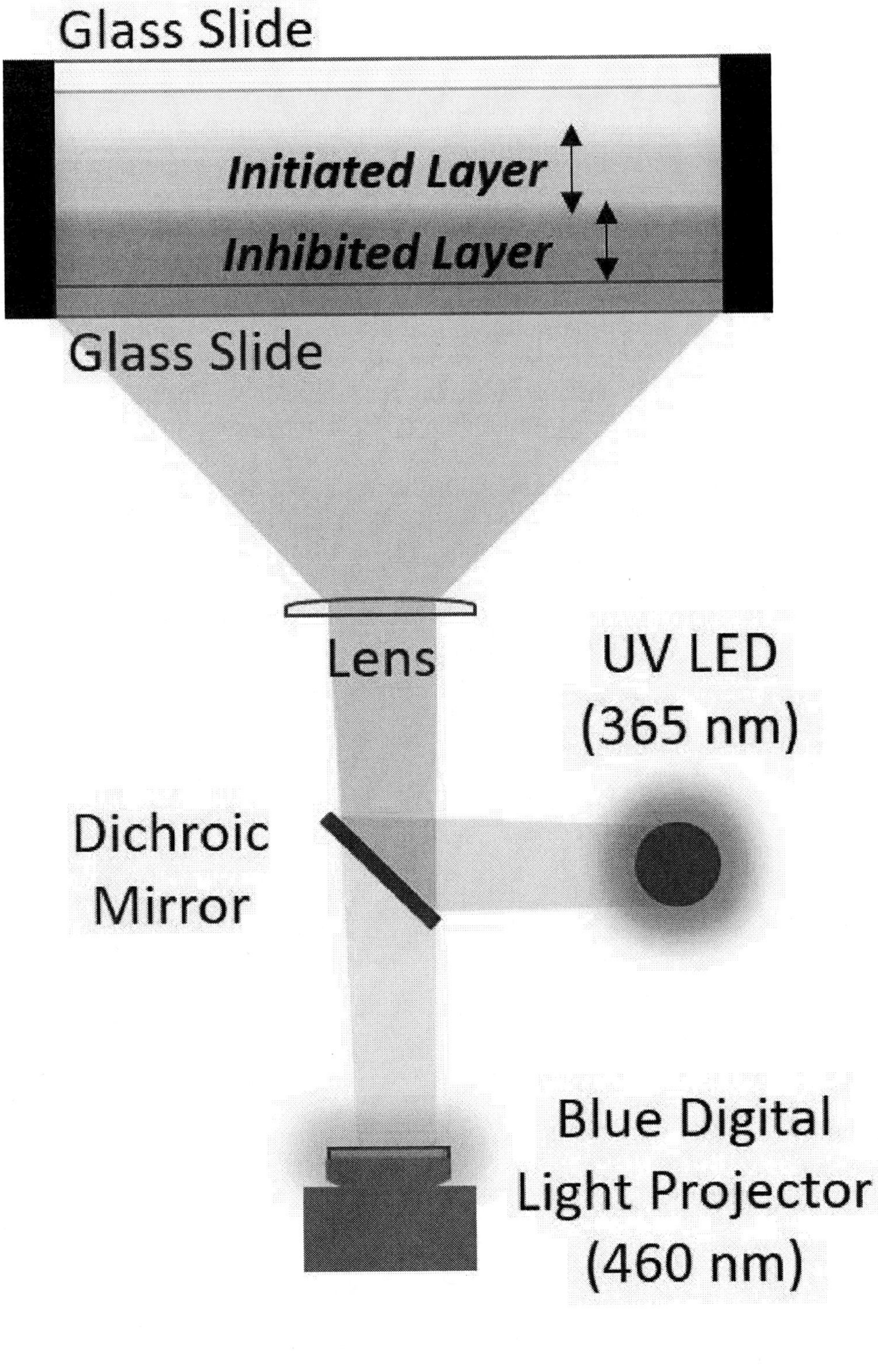
Glass Slide
Initiated Layer
Inhibited Layer
Glass Slide
Lens
UV LED
(365 nm)
Dichroic
Mirror
Blue Digital
Light Projector
(460 nm)

631075 Optical Interconnects on a Flexible Substrate Utilizing Additive Manufacturing Tools

Abstract:

The integration of flexible photonic components into flexible electronics provides new opportunities for applications in imaging, sensing, and optical interconnection between devices. While most photonic devices are currently printed on rigid substrates, their integration on flexible polymer substrates has also been demonstrated in recent years. However, current technology of contact and non-contact processes limits the type and performance of optical devices on these flexible substrates and requires special processes or product design to get the entry and exit of optical interconnect end facets out of the substrate plane. Here, we show that by utilizing recent advances in additive manufacturing, we can print photonic interconnects directly on flexible substrates.

631879 Cold Spray of Polystyrene Particles on Various Substrates

Abstract:

Cold spray (CS) is a solid-state coating technique that has been used since the 1980's for a wide range of applications for metals, polymers, ceramics, and composites on various substrates [1]. During the cold-spray process, particles are accelerated to high speeds through a converging and diverging nozzle with a pre-heated, high-pressure gas. Upon impact on a substrate, particles deposit to form a non-porous coating on the substrate that is caused by severe plastic deformation of the particle. The bond that forms between the particle and the substrate can be metallurgical, chemical, or mechanical. Cold spray of polymers has been investigated to a lesser extent than metals, and since polymers behave significantly differently than metals during the cold-spray process, many unknowns still exist [2–5]. Cold spraying of polymers have many current and potential applications, including surface protection from corrosion [6], protection from cavitation erosion or mechanical impacts [7], electronic applications [8, 9], and biocompatible and antibacterials coatings [10, 11]. However, the deposition efficiency of polymer cold spraying (10%) is very low when compared to metal cold spraying (~100%).

This study focuses on cold-spray deposition of polystyrene particles on polymeric and aluminum substrates. A number of different process conditions were investigated to achieve buildup of particles on the substrates. Analysis was performed to characterize the deposition thickness on each substrate. Promising results were obtained with polymeric substrates when the gas temperature and substrate heating was kept around the glass-transition temperature of the particle. As shown in figure, with 10 passes, 15 μm deposition was obtained at these conditions. The deposition on the aluminum substrates is still an ongoing investigation that is mainly focused on the first-layer adhesion, since the first couple of layers of deposited particles plays a key role during the process.

References

[1] Anatolii Papyrin, Vladimir Kosarev, Sergey Klinkov, Anatolii Alkhimov, and Vasily M Fomin. Cold spray technology. Elsevier, 2006.

[2] Abdulaziz S Alhulaifi, Gregory A Buck, and William J Arbegast. Numerical and experimental investigation of cold spray gas dynamic effects for polymer coating. Journal of Thermal Spray Technology, 21(5):852–862, 2012.

[3] Kesavan Ravi, Yuji Ichikawa, Tiana Deplancke, Kazuhiro Ogawa, Olivier Lame, and Jean-Yves Cavaille. De- velopment of ultra-high molecular weight polyethylene (uhmwpe) coating by cold spray technique. Journal of Thermal Spray Technology, 24(6):1015–1025, 2015.

[4] Y Xu and IM Hutchings. Cold spray deposition of thermoplastic powder. Surface and Coatings Technology, 201(6):3044–3050, 2006.

[5] Trenton B Bush, Zahra Khalkhali, Victor Champagne, David P Schmidt, and Jonathan P Rothstein. Optimiza- tion of cold spray deposition of high-density polyethylene powders. Journal of Thermal Spray Technology, 26(7):1548–1564, 2017.

[6] Srinivas Pravin Sitaram, James O Stoffer, and Thomas J O?Keefe. Application of conducting polymers in corrosion protection. Journal of Coatings Technology, 69(866):65–69, 1997.

[7] YQ Wang, Li P Huang, WL Liu, and J Li. The blast erosion behaviour of ultrahigh molecular weight polyethy- lene. Wear, 218(1):128–133, 1998.

[8] Lieh-Li Lin, Tsung-Han Ho, and Chun-Shan Wang. Synthesis of novel trifunctional epoxy resins and their modification with polydimethylsiloxane for electronic application. Polymer, 38(8), 1997.

[9] James J Licari. Coating materials for electronic applications: polymers, processing, reliability, testing. 2003.

[10] Anna M Vilardell, N Cinca, A Concustell, S Dosta, IG Cano, and JM Guilemany. Cold spray as an emerging technology for biocompatible and antibacterial coatings: state of art. Journal of Materials Science, 50(13):4441– 4462, 2015.

[11] Anke Zeller, Anna Musyanovych, Michael Kappl, Anitha Ethirajan, Martin Dass, Dilyana Markova, Markus Klapper, and Katharina Landfester. Nanostructured coatings by adhesion of phosphonated polystyrene particles onto titanium surface for implant material applications. ACS applied materials & interfaces, 2(8):2421–2428, 2010.

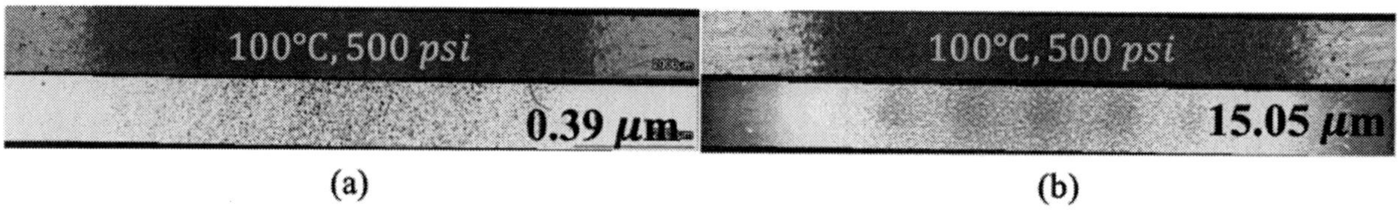

Figure. Comparison of PP surfaces at 10 passes (a) without and (b) with substrate heating at 100 °C with the corresponding thickness values.

630439 Use of Dimensionality Reduction and Transfer Learning in Deep Reinforcement Learning Controller for Hydraulic Fracturing

Abstract:

Reinforcement Learning (RL) is an area of machine learning where an agent learns optimal control actions by directly interacting with the environment in order to maximize the cumulative sum of rewards. However, the application of RL to process control has been recent and limited [1]. There are multiple approaches to solving the RL problem in the literature but recent successes in RL has been due to combining concepts from deep learning with RL algorithms. The idea of leveraging deep neural networks (DNNs) as function approximators in RL has delivered tremendous success in video games like Atari games [2] etc. On the contrary, in process control, a Deep RL (DRL) controller was recently proposed to control discrete-time nonlinear processes [3]. The DRL controller is a model-free off-policy actor-critic algorithm in which learning is based on temporal difference (TD) error [1] and Deterministic Policy Gradient (DPG) theorem [4]. The actor and the critic are represented using DNNs so that the policy and the value functions can be generalized to the continuous state and action spaces. Ideas like replay memory, target networks, and gradient clipping were used in the DRL controller to make learning appropriate for process control applications. The DRL controller was able to track setpoints for single-input-single-output (SISO), multi-input-multi-output (MIMO), and a nonlinear system with external disturbances [3].

Hydraulic fracturing is the extraction of oil and gas from rocks with low porosity and permeability. In order to extract oil and gas from such rocks, an artificial medium of proppant (sand) is created within the fractures using controlled explosions followed by the pumping of fracturing fluids at high pressures. The efficiency of the extraction process is dependent on the final fracture geometry and proppant concentration along the fracture. In order to achieve the desired objectives with respect to these process variables, model predictive controllers (MPC) have been designed recently [5]. But these controllers (a) require an accurate model which is difficult to obtain in the case of hydraulic fracturing given its complexity, (b) involve exorbitant computational costs, and (c) require regular re-tuning of controller parameters. Therefore, in this work, we propose to design a model-free DRL controller for hydraulic fracturing. Despite its success, the DRL controller has a few limitations which include the requirement of long training times, and careful initialization of hyperparameters for fast convergence [6]. In order to overcome these challenges, we propose to implement transfer learning wherein the controller learns a suboptimal policy offline using a data-based reduced-order model (ROM) before learning online from the actual process. In this work, we used a high-fidelity model of hydraulic fracturing process as a virtual numerical experiment of the actual process. Additionally, we used principal component analysis (PCA) to reduce the dimension of the RL state before using it for learning. Consequently, the actor and the critic were trained in the reduced-PCA space. With these proposed steps included in the DRL controller training, it shows convergence to an optimal policy with the objective of obtaining uniform proppant concentration along the fracture.

Literature cited:

[1] Sutton, R.S., Barto, A.G. Reinforcement Learning: An Introduction. *Cambridge: MIT Press*, 1998.

[2] Mnih, V., Kavukcuoglu, K., Silver, D., Graves, A., Antonoglou, I., Wierstra, D., Riedmiller, M. Playing Atari with Deep Reinforcement Learning. *arXiv Preprint, arXiv:13125602*, 2013.

[3] Spielberg, S., Tulsyan, A., Lawrence, N.P., Loewen, P.D., Bhushan Gopaluni, R. Toward self-driving processes: A deep reinforcement learning approach to control. *AIChE J*, **65**(10), 2019.

[4] Silver, D., Lever, G., Heess, N., Degris, T., Wierstra, D., Riedmiller, M. Deterministic policy gradient algorithms. *In ICML*, 2014.

[5] Siddhamshetty, P., Yang, S., Kwon, J. S.-I. Modeling of hydraulic fracturing and designing of online pumping schedules to achieve uniform proppant concentration in conventional oil reservoirs. *Comput. Chem. Eng.*, **114**:306 – 317, 2018.

[6] Shin, J., Badgwell T.A., Liu, KH., Lee, J.H. Reinforcement Learning – Overview of recent progress and implications for process control. *Computer Aided Chemical Engineering*, **44**:71-85, 2018.

632115 Harnessing Cognitive AI Technology in Refining to Enhance Operator Decisions

Abstract:

This talk will explore a case study on how one of the largest refiners in the US is leading the way to implement Cognitive AI technologies that digitize refinery campaign processes from planning to operator action. We'll explain how this technology is helping operators to choose a judgment about what action to take at any given time, leading to better coordination and higher refinery margin through intermediate upgrades, product giveaway, and unit optimization.

Digital technologies and AI are becoming crucial for refiners to improve margins and reduce costs. There is a desire to move quickly so that you can be the company that achieves higher profitability before it becomes the standard. Yet we are observing inefficiency and resistance to digital adoption. Resistance and inefficiency persist at the business level from uncertainly in how to define the most impactful use case, and at a personnel level from a lack of trust in novel digital technologies. In this talk, we'll discuss how operations teams can begin building trust in AI technologies and share a roadmap of how a refinery launched a major initiative to optimize refinery processes and improve profitability by digitizing the refinery operating plan and leveraging best practices and knowledge to enhance operator decision making abilities.

Commercial optimization of a refinery is a unique exercise in many respects, providing both challenge and opportunity. These include crude slates that change according to market demand, variability in feedstock quality, and the continually changing operating capability of equipment from degradation and many other factors. What makes one refinery perform better than its competitors is often its ability to execute a set of objectives consistently while navigating through this uncertainty. According to Accenture, 61% of refiners believe process optimization technologies are a primary benefit driver to production planning and execution. Refinery campaigns are fundamental to improving margins but using a traditional problem-solving workflow between engineers and operators can be ineffective resulting in a very transactional execution process, relying heavily on individual interpretation of the guidance. The burden is on the operator to fully understand context, leading to inconsistencies over time and between different teams.

To close the talk, we'll highlight primary benefits from Cognitive AI technologies that are being used in refining today, including providing context to operators, quantifying decision confidence, and allowing for accelerated improvement across campaign plans, ultimately closing the profitability gap.

3 main takeaways:

1. How refiners can build trust in novel digital technologies including artificial intelligence and define the more impactful use cases for their refinery.
2. How one of the largest refineries in the US implemented Cognitive AI technologies to optimize refinery processes and improve profitability by digitizing refinery campaign processes from planning to action in operations.
3. New insights into how Cognitive AI technologies are helping operators to make a judgment about what action to take at any given time, leading to better coordination and helping refineries to better capture the full refinery margin.

629822 Regression Model for Tool Wear Monitoring in Precision Machining

Abstract:

In metal cutting operations, tool wear must be monitored either by operators or by a tool wear monitoring system to avoid possible damage to products and machine tools. In precision machining specifically, the naturally occurring and progressing flank wear leads to unwanted vibration, which deteriorates the surface finish and results in dimensional inaccuracy of the machining part. In the worst case, the cutting tool can break while it is engaged with the workpiece. With the ever-increasing demand for cost savings and improvements in quality, tool condition monitoring is becoming increasingly important in modern manufacturing [1].

Modern tool wear monitoring systems leverage data-driven models applying machine learning (ML) to predict tool wear and detect machine abnormalities [2]. Many studies have involved ML algorithms that use techniques such as K-nearest neighbors [3], support vector machine [4], and convolutional neural networks [5]. Although these methods have shown good performance, the fundamental methodology and complete toolchain that is suitable to predict tool wear still need to be investigated. Moreover, regression models that predict tool wear as a continuous process state/variable are seldom. The often-applied classification aims to predict discrete class labels, but tool wear is a continuously progressing system disturbance. Therefore, it is important to explore regression models to predict tool wear of continuous variables. The objective of this work is to propose a regression-based framework for predicting tool wear in CNC systems. We use support vector regression (SVR) as the ML algorithm, to provide insights into and quantify tool wear prediction in the machining process.

As shown in Fig.1, in the proposed framework, we collect, tabulate, and label audio and vibration data of the CNC process under varying tool wear and explore the performance of the proposed framework using regression. First, audio and vibration signals are converted to frequency domain through Fast Fourier Transform (FFT) to reduce the complexity of the audio and vibration time waveforms collected. Then, Principal Components Analysis (PCA) is used to eliminate features of small variance, increase the information content, reduce overfitting and computational effort. The selected features of audio and vibration signals are modeled and analyzed with an SVR model. The framework of Fig.1 was applied to a Mazak Variaxis 630-5X II T Mill/Turn Machine 40 hp 5-axis CNC machine. The experimental facility was instrumented to acquire process data for machine power, spindle power, spindle vibration, and to record audio/video and is MTConnect capable. Moreover, the machine was controlled to perform milling at various spindle speeds, widths of cut, depths of cut, and feed rates. Application of the framework of Fig. 1 for tool wear monitoring will be presented. In terms of the accuracy in tool wear prediction, it will be shown that the audio and vibration signals are sufficiently rich in information about tool wear so that a very accurate model (0.013 and 0.007 in RMSE, respectively) could be developed with those signals. Computation of the relative error between real tool wear measurement and predicted tool wear showed 1.54% error in audio signal-based regression and 3.1% error in vibration signal-based regression. Fusion of these signals further improved the accuracy of the regression model. In conclusion, the acoustic signals that often lead engineers and operators identify issues and anomalies in the operation of precision machining equipment were used successfully to learn ML models to predict tool wear in a very accurate manner. Extrapolation of these models in time to be useful in machine prognostics is currently underway and will be discussed in this presentation.

Acknowledgment: This material is based upon work supported by the U.S. Department of Energy's Office of Energy Efficiency and Renewable Energy (EERE) under the Advanced Manufacturing Office Award Number DE-EE0007613.

makes any warranty, express or implied, or assumes any legal liability or responsibility for the accuracy, completeness, or usefulness of any information, apparatus, product, or process disclosed, or represents that its use would not infringe privately owned rights. Reference herein to any specific commercial product, process, or service by trade name, trademark, manufacturer, or otherwise does not necessarily constitute or imply its endorsement, recommendation, or favoring by the United States Government or any agency thereof. The views and opinions of authors expressed herein do not necessarily state or reflect those of the United States Government or any agency thereof."

References

[1] H. Shao, H. L. Wang, and X. M. Zhao, "A cutting power model for tool wear monitoring in milling," *Int. J. Mach. Tools Manuf.*, vol. 44, no. 14, 2004, doi: 10.1016/j.ijmachtools.2004.05.003.

[2] Z. Li, R. Liu, and D. Wu, "Data-driven smart manufacturing: Tool wear monitoring with audio signals and machine learning," *J. Manuf. Process.*, vol. 48, 2019, doi: 10.1016/j.jmapro.2019.10.020.

[3] A. Jegorowa, J. Górski, J. Kurek, and M. Kruk, "Use of nearest neighbors (K-NN) algorithm in tool condition identification in the case of drilling in melamine faced particleboard," *Maderas Cienc. y Tecnol.*, vol. 22, no. 2, 2020, doi: 10.4067/S0718-221X2020005000205.

[4] D. Shi and N. N. Gindy, "Tool wear predictive model based on least squares support vector machines," *Mech. Syst. Signal Process.*, vol. 21, no. 4, 2007, doi: 10.1016/j.ymssp.2006.07.016.

[5] F. Jia, Y. Lei, J. Lin, X. Zhou, and N. Lu, "Deep neural networks: A promising tool for fault characteristic mining and intelligent diagnosis of rotating machinery with massive data," *Mech. Syst. Signal Process.*, vol. 72–73, 2016, doi: 10.1016/j.ymssp.2015.10.025.

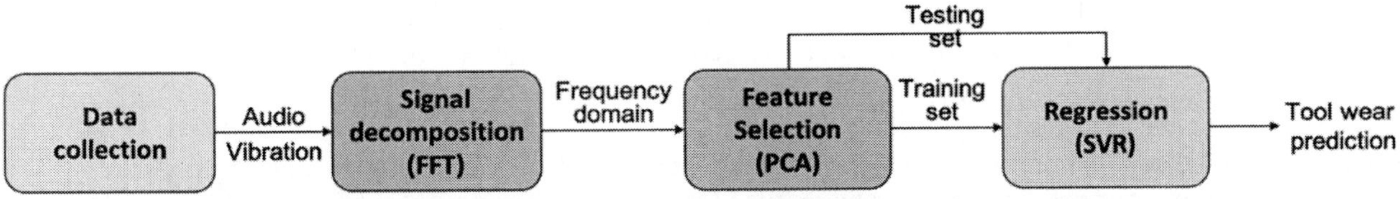

Figure 1. Flow diagram of a regression-based framework for tool wear prediction.

626215 Selection of Combined Index Weights to Optimize Anomaly Detection in Big Area Additive Manufacturing

Abstract:

As additive manufacturing (AM) becomes more prominent in our society, there are many questions about AM's repeatability and reliability. Today, AM is a process that involves constructing an object layer-by-layer; although this method allows the realization of complex computer-aided design (CAD) files into physical objects, a layer-by-layer process can also be more prone to error. Therefore, each layer has to be qualified so that the whole part can have a functional, rather than a purely aesthetic, purpose. Although AM has been used for several decades, most of the final part quality can be attributed to extensive operator knowledge and experience, rather than to the use of data-informed approaches. To assist operators, particularly those who may not be as knowledgeable about the process, and to monitor part quality, it is important to adapt data-driven approaches to warn operators of impending anomalies in the process. To achieve this, it is important to develop robust algorithms to interpret the vast quantities of collected data, especially when a priori knowledge of the data structure is nonexistent.

One traditional approach to monitor a manufacturing process is statistical process control (SPC). Before SPC is applied, a model that can adequately describe source of variation must be identified. Principal component analysis (PCA) [1] is one technique available to this end. The main idea is to describe normal variations by means of a number of principal components (PCs). This number (K) is smaller than the number of original variables (J). If there is no a priori knowledge about the data structure, it can be difficult to automatically calibrate a PCA model by selecting the correct number of PCs to retain. To overcome this issue, an existing proposed technique that involves the calculation of an ignorance score [2] to calibrate a probabilistic PCA model (PPCA) [3] is used in this study. An optimal model is selected by choosing the number of PCs for which the ignorance score achieves the minimal value.

After a model is selected using the ignorance score, SPC monitoring statistics such as Hotelling's T^2 statistic (H) and the squared prediction error (Q) can be used to detect anomalous data during process operation [4]. Anomalous data is identified when the H statistic and/or the Q statistic exceed their respective theoretical upper control limits (UCLs). While most applications make use of two charts, one for Q and one for H, a combined index () has been proposed more recently [5]. The combined index is a weighted sum of H and Q as seen in Eq. (1). The two weights, w_1 and w_2, are typically thought of as separate parameters [5]. So far, the weights of the combined index are calculated from UCLs for H and Q [5,6]. For example, the UCL of the H statistic relies on the assumption on how the H statistic is distributed (the chi-squared distribution was assumed in the original work [5]) and on selected confidence limits. Using UCLs as weights in the combined index is not motivated theoretically, only practically. Consequently, these UCL-based weights may render results obtained with the combined index difficult to translate from one application to the next. Today, there is no procedure available to set the weights (w_1 and w_2) so that optimal results can be assured, either theoretically or empirically. In this work, we present the first results of a systematic study where the chosen weights are varied deliberately in order to explore their effects on the combined index as a classifier during process monitoring.

Importantly, the combined index and its use for process monitoring is invariant to scale. Multiplying w_1 and w_2 by the same factor results in proportional change of the combined index and the upper control limit obtained for this combined index. To standardize our approach, we always chose the weights such that the sum of the eigenvalues of the empirical covariance matrix equals the following weighted sum of the same eigenvalues as seen in Eq. (2). The range of w_2 is chosen so that w_1 remains positive. Then, w_1 is calculated based on the above constraint.

Since the weights affect the estimated combined index, there likely exists an optimal set of relative weights that can optimize the performance of the combined index across numerous data types. The hypothesis is that the relative PPCA weights w_1=1 and w_2= (where) give the optimal anomaly detection performance. When these weights are applied, the combined index becomes equal to the squared Mahalanobis distance using the covariance matrix corresponding to the PPCA model [2,3] as seen in Eqs. (3-5). In essence, we are suggesting that the new optimal weights found in this study should be used consistently across literature to provide the best anomaly detection with the combined index.

The effect of the relative weights on the combined index is explored with simulated data sets [2]. Data types B1, B2, and C with noise to signal ratios up to 55% are used to generate simulated data sets with 1024 samples. The calculated combined indices are averaged over the number of samples. As seen in Figs. 1-2, the value of the combined index for a particular data set is a function of the selected weights. One should note that the vertical lines visually represent the locations of select weight ratios, such as the PPCA weights and UCL weights, where the chi-squared UCL has a specified confidence limit (i.e. 90%, 95%, and 99%). The simulated data sets allow the exploration of the effect of the noise to signal ratio and a nonoptimal number of retained PCs on the identification of the best performing relative weights. In order to compare the performance of each set of relative weights, the area under the curve (AUC) metric is used [7]. The AUC of the receiver operator characteristic curve (ROC) essentially describes the effectiveness of the combined index, given a set of relative weights, as a classifier to detect the true and false anomalies. The set of relative weights that lead to the maximum AUC (above 0.5 which is a random classifier) is the set of relative weights that produce the best classifier performance.

[1] Jolliffe, I.T. and Cadima, J., "Principal Component Analysis: A Review and Recent Developments." *Philosophical Transactions of the Royal Society A: Mathematical, Physical and Engineering Sciences*, 374(2065), 2016, pp. 20150202.

[2] Russo, S., Li, G., and Villez, K., "Automated Model Selection in Principal Component Analysis: A New Approach Based on the Cross-validated Ignorance Score," *Industrial & Engineering Chemistry Research,* 58(3), 2019, pp. 13448-13468.

[3] Tipping, M.E. and Bishop, C.M., "Probabilistic Principal Component Analysis," *Journal of the Royal Statistical Society: Series B (Statistical Methodology)*, 61(3), 1999, pp. 611-622.

[4] Montgomery, D., "Introduction to Statistical Quality Control", *Wiley*, 8th ed., 2019.

[5] Yue, H.H. and Qin, S.J., "Reconstruction-based Fault Identification Using a Combined Index," *Industrial & Engineering Chemistry Research*, 40(20), 2001, pp. 4403-4414.

[6] Jackson, J.E. and Mudholkar, G.S., "Control Procedures for Residuals Associated with Principal Component Analysis," *Technometrics*, 21(3), 1979, pp. 341-349.

[7] Fawcett, T., "An Introduction to ROC Analysis," *Pattern Recognition Letters,* 27(8), 2006, pp. 861-874.

$$\gamma = \frac{Q}{w_1} + \frac{H}{w_2} \tag{1}$$

where γ is the combined index, Q is the squared prediction error, H is Hotelling's T^2 statistic, and w_1 and w_2 are weights.

$$\sum_{j=1}^{J} \lambda_j = w_1 \sum_{j=1}^{K} \lambda_j + w_2 \sum_{j=K+1}^{J} \lambda_j = w_1 \sum_{j=1}^{K} \lambda_j + w_2\,(J-K) \tag{2}$$

where J represents the dimension of the data, K represents the number of retained PCs, $\sum_{j=1}^{K}\lambda_j$ is the sum of the retained eigenvalues, $\sum_{j=K+1}^{J}\lambda_j$ is the sum of the residual eigenvalues, and $\sum_{j=1}^{J}\lambda_j$ is the sum of all eigenvalues from the empirical covariance matrix.

$$\gamma_i = y_i^T(\Sigma)^{-1}y_i \tag{3}$$

$$\Sigma = w_1 \overline{P\Lambda P}^T + w_2 \underline{PP^T} = P\Lambda^* P^T \quad (4)$$

$$\Lambda^* = \begin{bmatrix} w_1\overline{\Lambda} & 0 \\ 0 & w_2 I \end{bmatrix} \quad (5)$$

where γ_i represents the combined index of the ith row, y_i represents the ith row in the mean centered data matrix (n samples x J variables), Σ is the covariance matrix corresponding to the PPCA model, $\overline{P}$ are the retained PCs (J x K), $\overline{\Lambda}$ is a diagonal matrix of retained eigenvalues (K x K), $\underline{P}$ are the residual PCs ((J-K) x K), P are all of the PCs (J x J), and Λ^* is a diagonal matrix of all of the weighted eigenvalues (J x J).

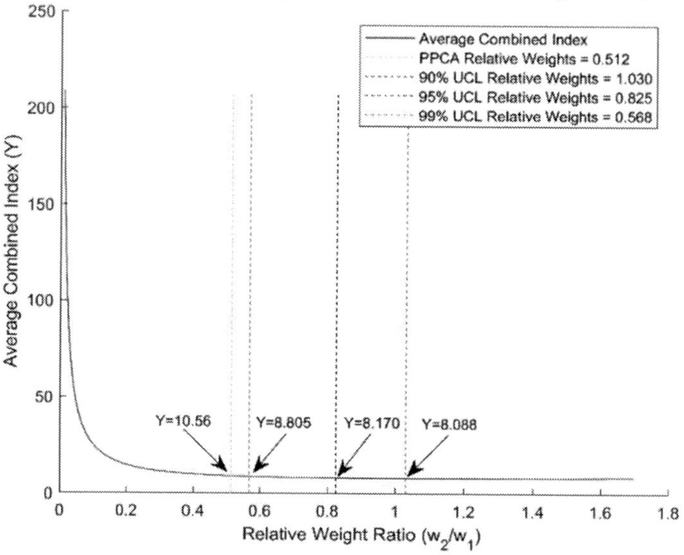

Figure 1. The combined index of data type B1.04 with a noise to signal ratio of 25% (calibrated with 3 PCs for a 9 dimensional data set). The vertical lines represent the locations of select relative weights. The theoretical UCL confidence level corresponds to the chi-squared control limit for the H statistic. The specified values on the plot represent the value of the combined index at the intersection of the selected relative weights indicated by the vertical lines.

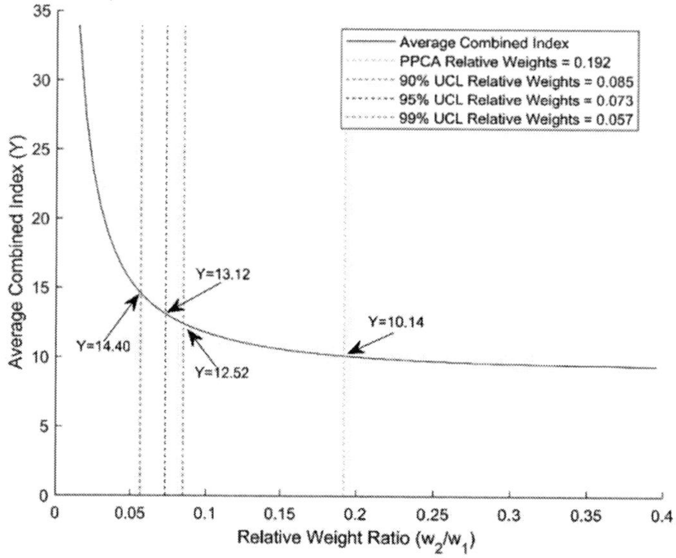

Figure 2. The combined index of data type C.02 with a noise to signal ratio of 25% (calibrated with 8 PCs for a 10 dimensional data set). The vertical lines represent the locations of select relative weights. The theoretical UCL confidence level corresponds to the chi-squared control limit for the H statistic. The specified values on the plot represent the value of the combined index at the intersection of the selected relative weights indicated by the vertical lines.

634105 Keynote Talk - Topological Data Analysis: Concepts, Computation, and Applications in Manufacturing

Abstract:

A key hypothesis that drives engineering practice is that data has structure. The dominant paradigms for describing such structure are statistics (e.g., moments, correlation functions) and signal processing (e.g., convolutional neural nets, Fourier series). Topological Data Analysis (TDA) is a field of mathematics that analyzes data from a fundamentally different perspective. TDA represents datasets as geometric objects and provides dimensionality reduction techniques that project such objects onto low-dimensional descriptors. The key properties of these descriptors (also known as topological features) are that they provide multiscale information and that they are stable under perturbations (e.g., noise, translation, and rotation). In this talk, we review key mathematical concepts and methods of TDA and present different applications in chemical engineering. Specifically, we discuss how to use TDA techniques to conduct real-time process monitoring, to design chemical sensors, and to quickly screen solvents for biomass processing.

634106 Keynote Talk - Platforms and Algorithms for Digitally-Enabled Next-Gen Manufacturing

Abstract:

Current initiatives calling for a significant increase in the degree of computing and communications networking technologies used to automate and optimize manufacturing can be traced to the 2011 – 2015 time period. The phrase "Industrie 4.0" was coined Wolfgang Wahlster and first used at the 2011 Hannover Fair in Germany (Kagermann, et al. (2011)). The phrase "Fourth Industrial Revolution" was defined in 2015 by Klaus Schwab (Schwab (2015)) and was used as the theme for the 2016 World Economic Forum Annual Meeting at Davos.

During the same time period, operating companies, notably ExxonMobil, issued a call for fundamental changes to the state-of-the-art of industrial control systems to address "pain points" including (1) a lack of interoperability between systems from different suppliers, (2) unacceptably high
costs for upgrades or replacements compared to other computing systems, and (3) restrictions on business value generation caused by barriers to insertion of third-party technology innovations (Bartusiak, et al. (2015)). This call to action resulted in the establishment of the Open Process Automation Forum of The Open Group in 2016 that is working to establish the industry standards for an open, secure, interoperable process automation architecture (https://www.opengroup.org/forum/open-process-automation-forum). The intent of the Open Process Automation Standard is to solve the identified industrial control system pain points and to enable realization of the vision of Industrie 4.0 and the Fourth Industrial Revolution for manufacturing.

Central to the Open Process Automation and Industrie 4.0 visions is a step-change increase in the pace of innovation with technologies including applied artificial intelligence, big data, and data analytics applied to manufacturing compared to the *status quo* with current industrial control system products that are of a closed, proprietary nature. Explicitly, one of the intentions of Open Process Automation is to increase opportunities for academic researchers and other technology providers to get their research transferred to industrial practice by reducing commercial barriers and the time it takes to deploy new technology.

In this talk, we will provide updates on the status of the Open Process Automation Standard (O-PAS), industrial prototypes of O-PAS based systems (Bartusiak, et al. (2021)), O-PAS conformance certification, and the state of Open Process Automation business ecosystem building. This will include guidance to academic researchers on how they can leverage the O-PAS Standard to facilitate the adoption of their research. We will comment on the relationships among Open Process Automation, Industrie 4.0, and similar initiatives -- such as the Clean Energy Smart Manufacturing Innovation Institute (CESMII) (Edgar and Pristikopoulis EN. (2018)), Manufacturing x Digital (MxD) (https://www.mxdusa.org), NAMUR Open Architecture (Klettner, et al. (2017)), and NAMUR Module Type Package (Bernshausen, et al. (2016)). Finally, we will comment on our outlook for new process systems engineering technologies -- including applied artificial intelligence, machine learning, big data, and data analytics -- to enable next-gen manufacturing efficiency.

References

Bartusiak RD, Bitar S, DeBari DL, Patin DL, Balsubramanian M, Foltz D, Caro RL, Brandl DL. (2015), "Functional Requirements for an Open Architecture Industrial Control System: Part 2 – What We Want". *ExxonMobil Research and Engineering Technical Report.*

Bartusiak RD, Bitar S, DeBari DL, Houk BG, Stevens D, Fitzpatrick B, Sloan P. (2021). "Open Process Automation: A standards-based, open, secure, interoperable process control architecture". *Proc. 2021 IFAC World Congress.*

Bernshausen J, Haller A, Holm, T, Hoernicke M, Obst, M, Ladiges J. (2016). "Namur Modul Type Package -- Definition". *atp-magazin* 58(1-2):72-81.

Edgar TF, Pistikopolous EN. (2018). "Clean Energy Smart Manufacturing Innovation Institute", Comp. Chem. Eng. 114:130-144.

Kagermann H, Lukas W-D, Wahlster W. (2011). "Industrie 4.0". VDI nachrichten. https://web.archive.org/web/20130304101009/http://www.vdi-nachrichten.com/artikel/Industrie-4-0-Mit-dem-Internet-der-Dinge-auf-dem-Weg-zur-4-industriellen-Revolution/52570/1

Klettner C, Tauchnitz T, Epple U, Northdurft L, Diedrich C, Schroeder T, Grossmann D, Banerjee S, Krauss M, Iatrou C, Urbas L. (2017). "Namur Open Architecture". *atp-magazin* 59(1-2):20-37.

Schwab K. (2015). "The Fourth Industrial Revolution". *Foreign Affairs*. https://www.weforum.org/agenda/2016/01/the-fourthindustrial-revolution-what-it-means-and-how-to-respond/

633993 Deep Composites: Bioinspired AI towards Modeling, Design and Manufacturing

Abstract:

Nature produces a variety of materials with many functions, often out of simple and abundant materials, and at low energy. Such systems - examples of which include silk, bone, nacre or diatoms - provide broad inspiration for engineering. Here we explore the translation of biological composites to engineering composites, using a variety of tools including molecular modeling, AI and machine learning, and experimental synthesis using 3D printing, and characterization. We review a series of studies focused on the mechanical behavior of materials, especially fracture, and how these phenomena can be modeled using a combination of molecular dynamics and machine learning. We present examples that involve deep convolutional neural networks, transformer neural networks, and game theoretical approaches towards design. One case study will cover a recent example that realizes a text-to-material design approach, developing new architected multimaterial composite designs based on human readable description and subsequent 3D printing – from word to matter. We conclude the talk with case studies of material optimization using genetic algorithms, applied to 3D printed composites, protein design, and a translation of protein folding to music and back to assess universal patterns.

Evaluation Of Process Variable Influence on Mechanical Properties of Short Fiber-Reinforced 3D-Printed Parts

Authors:

Martin Etemadi; Virginia Tech, Blacksburg, VA, USA
Arit Das; Virginia Tech, Blacksburg, VA, USA
Bradley A. Davis; Hume Center for National Security and Technology, Blacksburg, VA, USA
Steven H. McKnight; Virginia Tech Alexandria Campus (National Capital Region), Alexandria, VA, USA
Scott W. Case; Virginia Tech, Blacksburg, VA, USA
Christopher B. Williams; Virginia Tech, Blacksburg, VA, USA
Michael J. Bortner; Virginia Tech, Blacksburg, VA, USA

Abstract

Fused filament fabrication (FFF) has seen broad industrial adoption as it is capable of manufacturing large complex parts from robust thermoplastics in a cost-effective manner. However, the mechanical performance of the printed parts is limited due to poor interlayer bonding and the presence of voids. In order to overcome these drawbacks, the addition of short or continuous fibers into the polymer matrix has been investigated, as the fibers can act as a mechanical reinforcement while also mitigating residual stress resulting from the material's rapid solidification following extrusion. Therefore, understanding the implications of process parameters and fiber reinforcements on printed part properties through detailed crystallization analysis and rheological characterizations is of paramount importance.

The goal of this study is to understand the process–structure–property relationships of short carbon fiber reinforced polyamide 6 (CF-PA6) by comparing the melt rheology and crystallinity of CF-PA6 versus a neat PA6 polymer. Differences in the melting and crystallization behavior resulting from the reinforcing fibers revealed an increased time window for crystallization in the fiber-reinforced matrix. Rheological characterizations at the recommended printing temperatures demonstrate the shear-thinning behavior of the samples at shear rates relevant to FFF. From a statistical design of experiments analysis, the layer thickness was found to be the most significant parameter affecting the tensile properties of a printed part at a constant printing temperature and printing speed. The tensile fracture surfaces of the printed specimens using scanning electron microscopy were analyzed to provide insights into the failure mechanisms as a function of AM processing variables.

KEYWORDS:

3D printing, additive manufacturing, composites, polyamides, rheology

622245 Direct Ink Write of High Solid Suspensions: Considerations in Particle Type and Binder Properties

Abstract:

Construction, energetics, and regenerative medicine are several fields which rely on processing of suspensions with high content of solid particles (> 60 vol%). New processing methods for these areas, such as direct ink write (DIW) 3D printing, have enabled greater part complexity, reduced material waste, and improved safety during manufacturing. However, maximizing solid content presents processing difficulties due to the limited amount of binder in these suspensions, namely high viscosities, poor print resolution, and heterogeneous particle distributions. Despite these complications, for systems with particle diameters on the order of 10^2 μm, binder effects become more pronounced due to larger interstitial spaces between particles, but these effects remain poorly understood. Providing insights into the complex binder-particle and particle-particle interactions that occur at this scale is crucial for the development of successful DIW formulations. Here, we investigate the dual effects of particle type as well as binder factors in two different solidification mechanisms: solvent evaporation and UV cure. Using rheological tests, we characterize the viscoelastic response of these inks to applied stress mimicking the extrusion process. We use oscillatory time and 3 Interval Thixotropy experiments to analyze solidification post-printing as well as structure breakdown during shear followed by regeneration. We also visualize the microstructure prior to printing and after curing through scanning electron microscopy. For particles, we test spherical SiO_2, non-uniform SiO_2, and non-uniform basaltic soil particles. Binder factors of interest are surface tension of solvents and molecular weight and glass transition temperature of polymers. Fundamental understanding of the interplay between these components during 3D printing will enable a high level of customization of DIW inks for successful processing, enabling precise geometries tailored for novel buildings, safer energetics, or personalized medical devices.

632114 Dimensionalization of Two-Phase Newtonian/Non-Newtonian Flow Problems.

Abstract:

Non-Newtonian fluids are present in a large variety of everyday life applications including paints, foods, adhesives, inks, cements, slurries, polymers and gels. Computational fluid dynamics (CFD) simulations have been widely used to understand, optimize and design products and processes that involve such non-Newtonian fluid flows. Further complicating such challenges are cases wherein flows involve the interactions of multiple phases. Systematically, understanding complex flows is difficult since they generally involve multiple forces including inertial, viscous, gravitational, capillary forces. The present study suggests dimensionless forms for the momentum and continuity equations along with two-phase level-set and phase field equations for non-dimensionalizing two-phase, non-Newtonian fluid flow problems. The Reynolds number, Galilei's number and Capillary number arise naturally as scaling factors. Fluid dynamic (CFD) simulations using COMSOL Multiphysics® were used to illustrate comparative numerical analysis between dimensional and dimensionless equation forms for two-phase, non-Newtonian pipe flow and free-surface flow (slump-flow) examples. The effect of non-dimensionlization on computational time was demonstrated for different scale of free-surface flow problem. The importance of not satisfying simultaneously the Reynolds and Galilei's number similarity for scaling free surface-flow problems is also discussed. Finally, a scaling parameter was quantified and illustrated for both dimensionless and dimensional form of the free-surface flow (slump-flow) problem.

627307 Influence of Relative Humidity on the Spreadability and Triboelectric Properties of Powders in Additive Manufacturing Processes

Abstract:

Metallic and polymer powders are widely used in Additive Manufacturing (AM) processes, with for example Selective Laser Melting (SLM) and Selective Laser Sintering (SLS). During such operations, successive thin layers of powder are created with a ruler or with a rotating cylinder and then partially sintered with an energy beam. The layer thickness defines the vertical resolution, and the spatial homogeneity of the layer is a key parameter to ensure good and consistent quality of the built parts. Consequently, assessing the spreadability/flowability of powders is required to ensure good and consistent quality of the additively manufactured parts. The flowability of powders depends not only on their intrinsic properties (grain size, shape, etc.) but also on the measurement/processing conditions. Therefore, it is crucial to investigate the influence of the measurement conditions.

In this work, we investigated the influence of the relative humidity on the spreadability, packing dynamics and triboelectric behaviour of several common metallic powders (AlSi7Mg, Inconel, and SS316) used in additive manufacturing applications. The spreadability/flowability of the powders was characterized using the rotating drum method, to reproduce the recoating conditions (flow field and stress state) encountered in an additive manufacturing process. The packing dynamic was measured with the GranuPack instrument to study the bed density. Finally, the triboelectric behaviour of the powders was measured with a GranuCharge.

We show that the relative humidity modifies significantly the cohesiveness, the packing dynamics and the tendency to create electrostatic charges during a flow. The modification of the flow properties and triboelectric behaviour has a direct influence on the spreadability of powders in additive manufacturing applications. This study demonstrates the influence of the environmental conditions on the flow properties of powders used in additive manufacturing applications.

630064 Toward Next Level of Pharmaceutical 3D-Printing through Advanced Lipid-Based Excipients

Abstract:

The opportunities and challenges faced by three-dimensional (3D-) printing in the pharmaceutical field are not based solely on the process itself, but also associated with the material attributes. Until now, polymer-based excipients remain the principal materials employed for dosage formulation and printing of medicine, with very limited trials on lipid-based excipients (LBEs). The high amorphous content of polymers provides them with high flexibility which impacts their 3D printability. LBEs, unlike polymers, are characterized by a crystalline nature which contributes to their brittleness, making their 3D-printing very challenging. Considering the advantages of LBEs compared to polymers, e.g. being naturally occurring materials, predominantly digestible with "Generally Recognized as Safe" (GRAS) status, it is of significant importance to investigate their functionality for 3D-printing of pharmaceutical dosage forms. Recently, polyglycerol esters of fatty acids (PGFAs), also known commercially as Witepsol® PMF, were introduced as the next generation lipid-based excipients offering plenty of potentials for pharmaceutical applications [1]. The broad functionality based on their composition make PGFAs an appealing material for investigation. The synthesis of this group of excipients typically yields a mixture of PGFA as the main component (>80% w/w) and two more species, namely free polyglycerins (PGs) and mono acid fractions (MAF). Tuning the ratio between the three species can influence the material attributes and hence their processability and printability. In this study, we are investigating PGFAs, based on variations in their composition, as an advanced LBE for manufacturing lipid-based drug delivery systems via extrusion-based 3D-printing.

Due to the high number of polyglycerol moieties and free hydroxyl groups per molecule associated with the significant number of hydrogen bonds, and hence higher flexibility, the PG6-C16 partial ester (Witepsol PMF166) was selected as the starting material in this study. Beside the raw material, different blends of PG6-C16 partial esters composed of different ratios of its three components were prepared; either by adding PGs, MAF or both to the raw material. The materials were prepared for extrusion by melting, re-solidifying, crushing and sieving (<800 μm). Filaments of the PG6-C16 partial ester and its blends were manufactured using solid lipid extrusion at temperatures below the melting onset. The extrusion process parameters were adjusted in order to achieve 3D-printable filaments with targeted diameter and ovality using a 1.8 mm-diameter nozzle. By modifying the gear wheels of the printer, the force induced on the filament during 3D-printing was reduced to a minimum, enabling a flawless processability. A drug delivery matrix in the shape of a solid oral tablet ($10 \times 10 \times 4$ mm^3) was then printed using the produced filaments. In addition, extrudability and printability of other conventional lipids such as behenoyl polyoxyl-8 glycerides (containing PEG-8 esters) and polyethylene glycol monostearate (containing PEG-32 esters) were compared.

The printability was strongly impacted by the filament flexibility. The flexibility was assessed in terms of the critical bending diameter (CBD), which is the smallest bending diameter at which the material breaks under a bending load. The smaller the CBD, the more flexible is the filament. All LBEs were found extrudable, however, behenoyl polyoxyl-8 glycerides was not processable below its melting onset. The CBD of PG6-C16 partial ester, behenoyl polyoxyl-8 glycerides and polyethylene glycol monostearate were 100, 125, and >223 mm, respectively. The filaments of PG6-C16 partial ester were more flexible, hence easier to handle for printing than other tested lipids. Polyethylene glycol monostearate was unprintable with the present set-up. Varying the composition of PG6-C16 partial ester in terms of PGs and MAF, displayed upgraded performance in extrudability and printability. The highest flexibility was found at the highest ratio of PGs and MAF corresponding to a CBD of 1.8 mm (50 folds lower than that of PG6-C16 partial ester). Higher flexibility led to a more efficient process. The filaments were spoolable during extrusion and less prone to breakage, which, in turn, allowed simultaneous printing of multiple tablets.

In conclusion, PG6-C16 partial ester was effectively extruded and 3D-printed via tuning its composition. This approach is an essential step towards using LBEs as natural and biodegradable alternatives to polymer-based excipients, expanding the pallet of materials that can be used for 3D-printing of medicines. Nevertheless, more opportunities and new functionalities can be enabled through optimizing PGFAs properties.

References

[1] Corzo, C., Lopes, D. G., Lochmann, D., Reyer, S., Stehr, M., & Salar-Behzadi, S. (2020). Novel approach for overcoming the stability challenges of lipid-based excipients. Part 1: Screening of solid-state and physical properties of polyglycerol esters of fatty acids as advanced pharmaceutical excipients. *European journal of pharmaceutics and biopharmaceutics*, *148*, 134–147. https://doi.org/10.1016/j.ejpb.2020.01.012

Acknowledgments: The Austrian Funding Agency (FFG)

This work was funded through the Austrian COMET Program by the Austrian Federal Ministry of Transport, Innovation, and Technology (BMVIT), the Austrian Federal Ministry of Economy, Family, and Youth (BMWFJ), and by the State of Styria (Styrian Funding Agency SFG)

626851 Golem: A Probabilistic Approach to Robust Experiment and Process Optimization Based on Regression Trees

Abstract:

Numerous challenges in science and engineering can be framed as optimization tasks, including the maximization of reaction yields, the optimization of molecular and materials properties, and the fine-tuning of automated hardware protocols. Data-driven optimization strategies are often adopted to solve these tasks efficiently. Increasingly, these strategies are coupled with automated hardware to enable autonomous experimental platforms. The majority of the sequential model-based strategies used, however, do not consider robustness against the variability of experiment and process conditions. In fact, it is generally assumed that these parameters are exact and reproducible. Yet some experiments may have considerable noise associated with some of their conditions, and process parameters optimized under precise control may be applied in the future under variable operating conditions. In either scenario, the optimal solutions found might not be robust against input variability, affecting the reproducibility of results and returning suboptimal performance in practice. In this talk, I will introduce *Golem*, an algorithm that is agnostic to the choice of experiment planning strategy and that enables robust experiment and process optimization. *Golem* identifies optimal solutions that are robust to input uncertainty, thus ensuring the reproducible performance of optimized experimental protocols and processes. It can be used to analyze the robustness of past experiments, or to guide experiment planning algorithms toward robust solutions on the fly. We assess the performance and domain of applicability of *Golem* through extensive benchmark studies and demonstrate its practical relevance by optimizing an analytical chemistry protocol under the presence of significant noise in its experimental conditions.

Reference: Aldeghi, M., Häse, F., Hickman, R. J., Tamblyn, I. & Aspuru-Guzik, A. Golem: An algorithm for robust experiment and process optimization. *arXiv:2103.03716* (2021).

628299 Learning Based Scheduling of Industrial Hybrid Renewable Energy Systems

Abstract:

In order to reduce the global CO_2 emissions for low-carbon development while improving the efficiencies of operation, energy systems have started undergoing major transitions to facilitate the broad integration of distributed renewable energy technologies. The inclusion of renewable energy technologies to a typical fossil-fuel based energy source have gradually transformed the concept of centralized generation into distributed energy systems. The choice of optimal power contribution among the available hybrid sources often depends on several factors such as the time of use (TOU) price of the grid electricity, renewable energy generation forecast, electricity demand estimation etc.

Due to the high intermittent nature of renewable energy sources, the system complexity increases, thereby making the optimal decision making more challenging using the existing physics-based model techniques. The inevitable uncertainties related to energy sector such as climate changes, energy market fluctuations, and energy technology developments often demand complex physics-based model and excessive computational time especially for control applications. The physics-based energy scheduling approach generally solves an optimization problem in background to decide the optimal power mix among the available power sources. However, model dependencies, parameter uncertainties and multi-dimensional impact of the embedded uncertainties impose the decision making to be more problematic, especially when such an approach is extended to large-scale and complex energy systems.

The utilization of machine learning techniques in the energy sector has gradually attracted researchers' attention for the effective management of the aforementioned complexities due to their purely data-driven nature. Reinforcement learning (RL), which is a branch of machine learning, can adopt model-free approaches for scheduling under uncertainties by employing agents to participate in the process of interest. The RL agents make control decisions and maximize its total reward by interacting with the simulation environment and evaluating the series of actions taken.

In this study, a data-driven approach that adopts RL techniques to determine the schedule of power mix in a hybrid renewable energy system (HRES) is presented, with a case study on the highly energy intensive chlor-alkali process. This process involves electrolysis of sodium chloride solution to produce chlorine as the primary product together with hydrogen and sodium hydroxide (caustic soda) solution as the secondary products. Almost 50-70% of the overall cost of operating a chlor-alkali process comes from electricity usage, thereby making the process highly sensitive to electricity price trends. To combat the increased grid dependency of such a highly energy intensive process and to achieve profitability, a grid connected hybrid energy system comprising of solar PV panels, wind turbines and fuel cells are considered that supplies power to the chlor-alkali process.

We implement Proximal Policy Optimization (PPO), an Advantage Actor-Critic RL method that enable samples to be trained for multiple epochs of mini-batch updates, to effectively distribute energy among the various available sources without having access to the dynamics of the system. In our exclusive experiments, the RL based scheduling provides efficient operational strategies with saving in cost and show nearly on-par performance with the optimal solution. For stochastic tasks under uncertainties, our RL scheduling approach demonstrates strong superiority over the conventional optimization. This framework can generally be used in any data-driven scheduling problems taking the advantages of its model-free property, and can be generalized to multiple task scenarios via transfer learning.

634107 Keynote Talk: Modelling and Monitoring with Dynamic Auto-Regressive Latent Variable Methods

Abstract:

With the advent of Industry 4.0, industrial data can be collected at very high sampling rates, leading to more complicated relation between process and quality variables[1, 2]. To extract valuable information from these variables and avoid the curse of dimensionality, multivariate latent variable methods (LVM) are widely used, such as partial least squares (PLS)[3], canonical correlation analysis (CCA)[4] and latent variable regression (LVR)[5]. PLS and CCA are employed to construct the relations between process variables **X** and quality variables **Y** by maximizing their covariance and correlation respectively, while LVR aims to maximize the projection of **Y** in the latent space. Several variants of LVM are proposed to achieve better modeling and monitoring performance, including concurrent ones[6-8] and nonlinear versions[9, 10].

The aforementioned models obtain satisfactory modeling and monitoring performance; however, only static variations are considered in these models, and this static assumption constrains their applicability in actual dynamic systems. Several dynamic variants were proposed to extract temporal information from the data. For instance, augmented matrices were formed to include lagged process and quality variables in the work of Ku et al.[11] and Qin and McAvoy[12]. However, it is hard to interpret their latent structures because of the augmentation matrices. To address this issue, dynamic inner LVM was proposed to construct dynamic outer model by maximizing the relations between current samples and a linear combination of past samples, and a dynamic inner model that is consistent with the dynamic outer model[13-16]. Dynamic weighted variants were also proposed in [1, 17] to reduce the redundant information between adjacent historical samples by introducing basis functions. For aforementioned dynamic variants of LVM, only dynamic cross-correlations between the current quality variables and past process and quality variables are considered, leaving large portion of valuable quality information unexploited.

In this work, a series of dynamic auto regressive LVM (DALVM) algorithms are proposed to fully exploit the valuable information of both process and quality variables. DALVM aims to maximize the dynamic relations between current quality samples and past process and quality samples, and to achieve consistency, an auto-regressive exogenous (ARX) model is designed in the inner model. In DALVM, in addition to the auto-regressive relations of quality variables, the cross-correlations between **X** and **Y** are also constructed. Further, aiming for comprehensive modeling and monitoring, subsequent decompositions are proposed to decompose the process and quality variables concurrently into seven dynamic and static subspaces, and the corresponding monitoring scheme is designed for all subspaces. Two industrial dataset, Tennessee Eastman process and the three-phase flow facility, are used to illustrate the advantages of the proposed algorithms.

References

[1] Zhu, Q., 2020. Auto-regressive modeling with dynamic weighted canonical correlation analysis. Journal of Process Control, 95, pp.32-44.

[2] Xu, B. and Zhu, Q., 2020. Online quality-relevant monitoring with dynamic weighted partial least squares. Industrial & Engineering Chemistry Research, 59(48), pp.21124-21132.

[3] Geladi, P. and Kowalski, B.R., 1986. Partial least-squares regression: a tutorial. Analytica chimica acta, 185, pp.1-17.

[4] Hardoon, D.R., Szedmak, S. and Shawe-Taylor, J., 2004. Canonical correlation analysis: An overview with application to learning methods. Neural computation, 16(12), pp.2639-2664.

[5] Zhu, Q., 2020. Latent variable regression for supervised modeling and monitoring. IEEE/CAA Journal of Automatica Sinica, 7(3), pp.800-811.

[6] Qin, S.J. and Zheng, Y., 2013. Quality-relevant and process-relevant fault monitoring with concurrent projection to latent structures. AIChE Journal, 59(2), pp.496-504.

[7] Zhu, Q., Liu, Q. and Qin, S.J., 2017. Concurrent quality and process monitoring with canonical correlation analysis. Journal of Process Control, 60, pp.95-103.

[8] Zhang, H. and Zhu, Q., 2021. Multi-layer fault monitoring with concurrent kernel latent variable regression. Submitted to Industrial & Engineering Chemistry Research.

[9] Melzer, T., Reiter, M. and Bischof, H., 2003. Appearance models based on kernel canonical correlation analysis. Pattern recognition, 36(9), pp.1961-1971.

[10] Zhang, Y., Zhou, H., Qin, S.J. and Chai, T., 2009. Decentralized fault diagnosis of large-scale processes using multiblock kernel partial least squares. IEEE Transactions on Industrial Informatics, 6(1), pp.3-10.

[11] Ku, W., Storer, R.H. and Georgakis, C., 1995. Disturbance detection and isolation by dynamic principal component analysis. Chemometrics and intelligent laboratory systems, 30(1), pp.179-196.

[12] Qin, S.J. and McAvoy, T.J., 1996. Nonlinear FIR modeling via a neural net PLS approach. Computers & chemical engineering, 20(2), pp.147-159.

[13] Dong, Y. and Qin, S.J., 2018. Dynamic latent variable analytics for process operations and control. Computers & Chemical Engineering, 114, pp.69-80.

[14] Dong, Y., Liu, Y. and Qin, S.J., 2019. Efficient dynamic latent variable analysis for high-dimensional time series data. IEEE Transactions on Industrial Informatics, 16(6), pp.4068-4076.

[15] Zhu, Q., Qin, S.J. and Dong, Y., 2020. Dynamic latent variable regression for inferential sensor modeling and monitoring. Computers & Chemical Engineering, 137, p.106809.

[16] Qin, S.J., Dong, Y., Zhu, Q., Wang, J. and Liu, Q., 2020. Bridging systems theory and data science: A unifying review of dynamic latent variable analytics and process monitoring. Annual Reviews in Control.

[17] Xu, B. and Zhu, Q., 2020. Online Quality-Relevant Monitoring with Dynamic Weighted Partial Least Squares. Industrial & Engineering Chemistry Research, 59(48), pp.21124-21132.

630085 Personalizing Wound Dressings Via 3D Printing for the Treatment of Thermal Burns

Abstract:

Burns are a common injury sustained worldwide, with an estimated 180,000 deaths caused annually according to the World Health Organization. Depending on the depth and area, burns can range from a trivial injury that heals in a few days to life-threatening wounds that is further complicated by the risk of bacterial infections, which may result in issues such as debilitating scarring upon recovery if not treated properly. While skin grafting remains the gold standard for the treatment of burn wounds and exist as various forms (eg: autologous skin grafting, allogenic skin grafting and cultured epithelial autografts), there is still a need for wound dressings to either complement skin grafting or to serve as an alternative in situations whereby skin grafting is not feasible. With the rise of 3D printing in recent times comes the possibility of personalized medicine that tailors drug dosages according to the patient's needs. Thus, the aim of our study is to incorporate this concept into wound dressings. Using a 3D printer with multiple printheads, we have fabricated hydrogel wound dressings with various shapes, sizes, architectures and designs through the subsequent deposition of different drug-hydrogel formulations and have observed how they affect the cumulative release profile of drugs embedded inside the dressing. This is beneficial as the dressing design can be tailored to accommodate the patient's condition and it allows for decentralized production that can be controlled by the physician to match their diagnosis.

631699 3D Printing of Pharmaceuticals-Exploring Process Parameters and Structure Dry Weight Relations

Abstract:

Three-dimensional (3D) printing technology has added new dimensions to pharmaceutical drug product manufacturing, such as designing dosage forms with tunable drug loading, delivering pharmaceuticals on-demand, creating multi-drug pills, etc. [1, 2]. Among several 3D printing processes, micro-extrusion-based pressure-assisted microsyringe (PAM) printing is popular due to its materials (i.e., polymers, drugs, functional excipients, etc.) handling capability at low temperatures. This research explores the relationship between key process parameters and the dry weight of the 3D printed structure (i.e., printlet or tablet) to identify the optimum printing conditions. A 3-compartment tablet design was designed and developed as a dosage form considering the future application of drug tunability and multi-drug printing capability. Sodium alginate-based formulation with fenofibrate as a drug or active pharmaceutical ingredient (API) was developed and used to print. Cellink BioX 3D printer was used to print. Printing speed, printing pressure, and infill density were identified as the key process parameters and chosen as independent variables in the Box-Behnken design, a response surface methodology (RSM). The shape fidelity ranking and dry tablet weight were chosen as the dependent variables. The design was applied with 15 experiments, including three center points. Each unique experiment was performed in triplicate, making a total of 45 experimental runs. The statistical analysis and surface plots showed that pressure and printing speed have an impact compared to infill density on the dry tablet weight. Optimal values for the pressure and print speed were found to be 60 kPa and 10 mm/s, respectively. Characterization (DSC, SEM, and assay) of the tablets confirm fenofibrate crystallinity, no imperfection of the 3D printed structure, and fenofibrate content of approximately 30%, respectively.

References:

[1] M. Azad, D. Olawuni, G. Kimbell, A. Badruddoza, Md. Hossain, T. Sultana, Polymers for Extrusion-based 3D Printing of Pharmaceuticals: A Holistic Materials-Process Perspective, (review paper) Pharmaceutics, 12 (124) (2020), 1-34.

[2] I. Fleming, I. Jama, J. Chikha, D. Olawuni, G. Kimbell, M. Azad, Design & Development of 3D Printed Polymeric Structure for Pharmaceutical(s) Delivery, AIChE Annual Meeting, Abstract ID 585668, Orlando, FL, Nov. 2019.

624109 Novel Polyester-Based Thermoplastic Elastomers for 3D-Printed Personalised Urethra Pessaries

Abstract:

Stress urinary incontinence (SUI) is the most prevalent type of incontinence with up to 46% of women being affected worldwide [1]. One effective means of treatment is the use of a urethra pessary (UP) – a medical device administered and shaped similarly to an intra-vaginal ring (IVR). UPs support the urethrovesical junction and alter the position of the urethra by mechanical stabilisation and therefore prevent involuntary urine loss [2]. The success rate of this type of therapy is highly dependent on a proper pessary fitting, though. Since commercially available UPs are only available in a handful of sizes, their success rate to successfully treat SUI is less than 50% [3]. The present study aims at tackling this shortcoming by presenting personalised UPs processed via material extrusion-based additive manufacturing, also known as fused filament fabrication or 3D-printing, for the first time. A polyester-based thermoplastic elastomer was employed as the raw material, which reveals comparable mechanical properties to the commercially employed silicones, satisfies the necessary regulatory requirements for medical devices as well as exhibits outstanding processability. As methods to assess the mechanical properties of UPs are scarce, the tensile and compression test set-ups commonly employed for IVRs were adapted to accurately capture their critical mechanical properties. By systematically varying the pessaries' outer (cross-sectional diameter) and inner geometry (infill percentage), a wide range of tensile and compression properties was achieved. All 3D-printed UPs covered a maximum tensile force above a critical female pull force, allowing a flawless pessary removal. Both the maximum compression force at a given deformation, which represents the ability to change the urethra position, and the component compression stiffness, which describes the resistance of the UP to deform under manual compression during pessary insertion, were strongly influenced by the choice of process parameters. Therefore, depending on the severity of SUI, the relevant mechanical properties of the 3D-printed UPs can be customised in a way that they either fall below (for mild cases of SUI), equal or exceed (for severe cases of SUI) those of the commercially available silicone UP with the same outer diameter. Neither an application-relevant long-term immersion into simplified simulated vaginal fluid, nor a long-term static pre-compression emulating the application loading, nor a combination thereof negatively impacted any of the described mechanical parameters. Consequently, the present study confirms the possibility to personalise the geometry as well as the mechanical properties of UPs via 3D-printing a polyester-based thermoplastic elastomer and, hence, to considerably increase the patience adherence and the success rate of UP to successfully and more effectively treat SUI.

1. Reynolds, W.S.; Dmochowski, R.R.; Penson, D.F. Epidemiology of stress urinary incontinence in women. *Curr. Urol. Rep.* **2011**, *12*, 370–376.

2. Al-Shaikh, G.; Syed, S.; Osman, S.; Bogis, A.; Al-Badr, A. Pessary use in stress urinary incontinence: a review of advantages, complications, patient satisfaction, and quality of life. *Int. J. Womens. Health* **2018**, *10*, 195–201.

3. Jones, K.A.; Harmanli, O. Pessary Use in Pelvic Organ Prolapse and Urinary Incontinence. *Rev. Obstet. Gynecol.* **2010**, *3*, 3–9.

623905 3D Printing of Chiral Liquid Crystal Elastomers Using Cellulose Nanocrystals

Abstract:

Cellulose nanocrystals, CNCs in polar solvents (such as water and DMSO) can self-assemble into a chiral nematic structure at low concentrations (below 10 wt.%). Therefore, the pure CNC suspensions do not have sufficient viscoelasticity required for 3D-printing applications. To enhance the viscoelasticity of the pure CNC suspensions in DMSO, we have incorporated photo-curable monomers (such as 2-Hydroxyethyl methacrylate (HEMA)), which improve the filament mechanical properties when exposed to UV light. We have studied the effect of the CNC:HEMA ratio in DMSO on the chiral nematic self-assembly of CNCs before and after 3D-printing. Our polarized optical microscopy results showed that the CNC:HEMA (wt%) ratios of 1:6 and 1:8 exhibit cholesteric structures at stationary conditions. We have performed flow-induced birefringence measurements to explore the effect of monomer addition on the flow-induced alignment. We have further studied the relaxation dynamics of the chiral structures after stopping the flow. Our results showed that, over the course of 3D printing, the shear flow in the 3D-printer nozzle can unwind the chiral nematic structure and align the CNC particles along the flow axis. The chiral nematic structure was, however, fully recovered a few minutes after stopping the flow. After printing, the chiral nematic ink (CNC+monomer) was exposed to UV light to complete the photopolymerization. The presence of chiral structures in the printed part was confirmed using Polarized Optical Microscopy and Scanning Electron Microscopy. Moreover, they reveal the perpendicular alignment of the chiral nematic structure with respect to the printing direction. These new findings can pay the way to develop new ink formulations for 3D printing of architectures with internal chirality for advanced mechanical and smart photonic applications.

632697 Control Techniques for Handling Sensor and Actuator Cyberattacks on Evolving Nonlinear Process Systems

Abstract:

Cyberattacks on control systems are a concern due to the possible catastrophic consequences of a successful attack. Despite various efforts in recent years to develop techniques for detecting attacks via machine learning [3] or state prediction [4], or handling attacks via other techniques (e.g., signal injection [5]), many open challenges with respect to control system cybersecurity remain. One of these challenges is that cyberattacks might be identified by evaluating deviations from "normal" process behavior. However, as the process dynamics change over time for a process, it could be that metrics designed to detect deviations from "normalcy" begin to flag dynamics changes as attacks. This concept motivated our work in [6], in which we extended two detection strategies from [7] that are integrated with the use of an optimization-based control strategy termed Lyapunov-based economic model predictive control (LEMPC) [1, 2] to the case where process dynamics change over time. Specifically, when there are no attacks or changes in the process dynamics, the closed-loop state is maintained within a characterizable region of operation at all times, and a threshold on either the difference between state predictions and measurements or between different state estimates is not breached. After either the closed-loop state leaves the characterizable region of operation or the difference between the predictions/measurements or estimates is breached, the threshold on the predictions/measurements or estimates is modified and model re-identification is triggered after a defined number of sampling periods. This strategy is able to theoretically guarantee that the closed-loop state is maintained within a characterizable operating region for at least one sampling period after an undetected attack or until the time the model re-identification occurs when there is only a model change and no attack.

Despite the elegance of the theories developed in [6], an important question to address is how such a strategy might be implemented practically. Specifically, the theoretical results in [6] introduce a number of theoretical expressions which must be satisfied, including, for example, various parameters related to the operating region and plant/model mismatch. Through a series of simulation examples involving a continuous stirred tank reactor and focused on the state estimation-based attack detection strategy, we elucidate some of the difficulties in designing resilient control strategies without rigorously determining the parameters of the control and detection strategies using theory but instead attempting to utilize simulation. We also discuss, via simulation, strategies for attempting to thwart specific types of attacks on the actuators via injecting control signals at intervals, as well as the role of plant/model mismatch introduced via data-driven models and numerical error on control actions when state measurement cyberattacks occur. Finally, in light of these control/detection concepts, we discuss concepts related to cyberattack discoverability for nonlinear systems.

References:

[1] Heidarinejad, M., Liu, J., & Christofides, P. D. (2012). Economic model predictive control of nonlinear process systems using Lyapunov techniques. AIChE Journal, 58(3), 855-870.

[2] Ellis, M., Durand, H., & Christofides, P. D. (2014). A tutorial review of economic model predictive control methods. Journal of Process Control, 24(8), 1156-1178.

[3] Wu, Z., Albalawi, F., Zhang, J., Zhang, Z., Durand, H., & Christofides, P. D. (2018). Detecting

and handling cyber-attacks in model predictive control of chemical processes. Mathematics, 6(10), 173.

[4] Cárdenas, A. A., Amin, S., Lin, Z. S., Huang, Y. L., Huang, C. Y., & Sastry, S. (2011, March). Attacks against process control systems: risk assessment, detection, and response. In Proceedings of the 6th ACM Symposium on Information, Computer and Communications Security, 355-366.

[5] Satchidanandan, B., & Kumar, P. R. (2016). Dynamic watermarking: Active defense of networked cyber–physical systems. Proceedings of the IEEE, 105(2), 219-240.

[6] Rangan, K. K., Oyama, H., & Durand, H. (2021). Integrated cyberattack detection and handling for nonlinear systems with evolving process dynamics under Lyapunov-based economic model predictive control. *Chemical Engineering Research and Design.*

[7] Oyama, H., & Durand, H. (2020). Integrated cyberattack detection and resilient control strategies using Lyapunov-based economic model predictive control. *AIChE Journal*, *66*(12), e17084.

623249 An Algorithm for Exascale-Capable Integrated Process Design and Control

Abstract:

Industrial processes are subject to endogenous and exogenous variability that has detrimental effects on their performance. It is therefore necessary to design process systems that inherently exhibit low sensitivity to variability, but also facilitate the adoption of operating strategies and mechanisms (e.g., controllers) that efficiently alleviate the detrimental effects of variability. These requirements are addressed through methods for integrated process design and control[1], which aim to design processes that are both economically optimum at steady state and robust under the influence of disturbances. Such methods need to account for a very wide range of process structural and operating options, simultaneously with numerous variability scenarios and detailed process models. The simultaneous consideration of all these features requires extremely intense computational effort.

Modern high-performance computing technologies in the form of upcoming exascale systems provide a unique opportunity to address the computationally demanding calculations. Exascale class machines will see a massive increase in the number of computing units (between tens and hundreds of millions), in the form of homogeneous cores or heterogeneous mixtures of multipurpose CPUs, GPUs and other specialized processing units[2]. However, the exploitation of such technologies includes significant challenges as they require the development of algorithms that exhibit high scalability, enable effortless portability within heterogeneous computing resources and support resiliency to failures. Available algorithms for integrated process design and control have not been developed with such capabilities. There are few occasions where such algorithms have been used in parallel environments, without specific references to optimizations that facilitate parallel execution[3]. Specific developments in optimization algorithms[4] and in model predictive control[5] for parallel environments are available, although the two are not combined and the latter are rare. Reduced-order modelling for process design and control is becoming increasingly popular[6], but the focus is not on optimization for parallel infrastructures. Optimization algorithms have been reported that make use of either CPUs or GPUs[4,5], but not both. Frameworks such as OpenCL are generally available for cross-platform, parallel programming[7], but specialized codes are needed depending on the type of computing unit utilized. Derivative-free optimization (DFO) algorithms often result in redundant computations as they lack the mechanisms to avoid them, whereas the challenges exhibited in the convergence of derivative-based optimization (DBO) algorithms is an open research topic. Algorithms with such features would require significant upgrading for use in parallel computing infrastructures, to avoid over- or under-utilization of computational resources and load imbalances, to improve their scalability and to enable exploitation of heterogeneous resources.

This work proposes a novel and generic algorithmic scheme for simultaneous process design and controllability assessment that combines approximate computing techniques with skeleton programming and run-time scheduling over heterogeneous computing nodes. In approximate computing techniques, exact computations are replaced by selective approximations that incur significant reduction in computational resources utilization. This comes at the expense of a small deterioration in the accuracy of the obtained solutions[8]. The algorithm is further developed using general purpose skeletons[9] for parallel matrix operations simultaneously on both CPUs and GPUs. Skeleton programming increases the developer's productivity by abstracting away parallelism, and enables effortless deployment on large scale computing systems. Finally, a run-time scheduling library[10] is employed that is also suitable for heterogeneous computing nodes and enables efficient load management.

These methods are incorporated into a hybrid scheme that comprises an external, DFO-based layer that handles discrete optimization variables and an internal, DBO-based layer for continuous process optimization and controllability assessment. In every iteration, the external DFO generates values for the discrete variables which are used by a DBO for steady-state process optimization using an economic objective function. The control performance of each optimum process design, generated by the DBO, is assessed by a homotopy-continuation algorithm. The latter is used to evaluate the non-linear sensitivity of the process in the context of a control structure, within a wide range of variability scenarios. The control structure takes the form of an objective function that accounts for the distance of each design solution under variability from the desired set-point and for the cost of the resources needed to bring the process back to its set-point. These two objective functions are aggregated into one that is used by the DFO to guide the overall optimization search until convergence. The employed approximate computing techniques include memoization, task dropping and loop perforation. Memoization is used at the level of the DFO to generate a record of previously visited solutions and to avoid their time-consuming re-evaluation in the internal layer. Task dropping is used within the DBO and the homotopy-continuation algorithms to efficiently avoid the time- consuming simulations that eventually result in solutions that fail to meet the algorithmic convergence criteria. Loop-perforation is used within the control performance assessment, in a scheme that enables gradual increase of the intensity of the investigated variability scenarios as the DFO proceeds to convergence. The skeleton programing and the run-time scheduling frameworks are used to parallelize the equality and inequality constraints that form the process model used by the DBO and the homotopy-continuation algorithms. Their implementation is adapted to the modular formulation of the superstructure used to develop the process model, enabling the simultaneous distribution of different process modules on the desired computing resources (CPUs or GPUs).

The proposed algorithm is implemented on the optimum design and control of chemisorption-based CO_2 capture flowsheets. A superstructure of absorption/desorption flowsheets is used that enables the consideration of wide stream distribution options, together with sizing of equipment and operating optimization[11]. Variability is considered in feed compositions, operating temperatures etc.[12]. The employed DFO is Simulated Annealing, whereas the DBO is the Interior Point Optimizer (IPOPT) and the homotopy-continuation algorithm is PITCON. Table 1 below shows indicative results from the implementation of memoization and task-dropping. It is clear that the number of function evaluations in each thread is much smaller due to the two techniques. The total CPU time to convergence of the new algorithm is 60 times lower. The optimum objective function value attained by the new algorithm is 44% better. These CPU time and objective function values correspond to a design space that is 2 orders of magnitude wider than the conventional algorithm, in terms of discrete parameter combinations.

Furthermore, the task dropping approach reduces on average by 63% the number of function evaluations that result in invalid simulations and this greatly improves the time performance of the algorithm. Further investigations are performed for the case of loop perforation, which exhibits similar improvements as the ones reported above. The skeleton programming and the scheduling frameworks further enable approximately 40% lower CPU time compared to an algorithm without them. Results will also be reported regarding the simultaneous use of CPUs and GPUs and the implementation in a supercomputer using up to 1000 parallel threads.

Acknowledgements

This work has received funding from the European Union's Horizon 2020 research and innovation programme, under grant agreement No. 801015 (EXA2PRO, https://exa2pro.eu/). This work was supported by computational time granted from the National Infrastructures for Research and

Technology S.A. (GRNET) in the National HPC facility - ARIS – under project EXACO2.

References

1. P. Vega, R. Lamanna de Rocco, S. Revollar and M. Francisco, *Comput. Chem. Eng.*, 2014, 71, 602–617.

2. S. Ashby, P. Beckman, J. Chen, P. Colella, B. Collins, D. Crawford, J. Dongarra, K. Kothe, R. Lusk, P. Messina, T. Mezzacappa, P. Moin, M. Norman, R. Rosner, V. Sarkar, A. Siegel, F. Streitz, A. White and M. Wright, *The Opportunities and Challenges of Exascale Computing, Summary Report of the Advanced Scientific Computing Advisory Committee (ASCAC) Subcommittee*, 2010.

3. D. C. Miller, D. Agarwal, D. Bhattacharyya, J. Boverhof, Y. Chen, J. Eslick, J. Leek, J. Ma, P. Mahapatra, B. Ng, N. V. Sahinidis, C. Tong and S. E. Zitney, in *Process Systems and Materials for CO 2 Capture* , 2017.

4. B. Sauk, N. Ploskas and N. Sahinidis, *Optim. Methods Softw.*, 2020, **35**, 638–660.

5. N. F. Gade-Nielsen, Interior Point Methods on GPU with application to Model Predictive Control, PhD *Thesis, Denmark Technical University*

6. J. H. Lee, J. Shin and M. J. Realff, *Comput. Chem. Eng.*, 2018, **114**, 111–121.

7. D. Grewe and M. F. P. O'Boyle, in *Lecture Notes in Computer Science (including subseries Lecture Notes in Artificial Intelligence and Lecture Notes in Bioinformatics)*, 2011, vol. 6601 LNCS, pp. 286–305.

8. B. Grigorian and G. Reinman, in *ACM Transactions on Architecture and Code Optimization*, 2015, vol. 12.

9. A. Ernstsson and C. Kessler, in *Parallel Computing: Technology Trends, series: Advances in Parallel Computing,* 2020, pp. 475–484.

10. C. Augonnet, S. Thibault, R. Namyst and P. A. Wacrenier, in *Concurrency Computation Practice and Experience*, 2011, vol. 23, pp. 187–198.

11. T. Damartzis, A. I. Papadopoulos and P. Seferlis, *Clean Technol. Environ. Policy*, 2014, **16**, 1363–1380.

12. P. Seferlis and J. Grievink, *Comput. Chem. Eng.*, 2004, **17**, 326–351.

Table 1: Results of novel algorithm with task dropping and memoization compared to the conventional parallelization approach.

	Conventional parallelization	Task Dropping + Memoization	Improvement (%)
Number of threads used	56	56	-
Number of function evaluations per thread	40,697	5,388	86
CPU time (sec)	777,600 (216 hr)	13,181 (3.5 hr)	98
Optimum objective value	20.79	11.67	44

632301 Integrated Actuator Attack Detection and Control for Nonlinear Systems Under Lyapunov-Based Economic Model Predictive Control

Abstract:

With the impetus that Industry 4.0 [1, 2] has received, especially over the last decade, many industries are moving toward greater integration of physical processes and sensor measurements with computer and communication systems. While this development of integrated cyber-physical systems (CPS) has advantages like increased production efficiency and transparency of the various steps involved in production, it also makes the processes vulnerable to malicious cyberattacks. Cyberattacks can cause major issues with guaranteeing safety of process systems and have hence motivated multiple projects which account for attacks on process sensors [4, 5, 6] and process actuators [7, 8]. Recent work in our group has developed strategies for handling cyberattacks on process sensor measurements with theoretical guarantees for nonlinear dynamic systems in the context of an advanced optimization-based control framework known as Lyapunov-based economic model predictive control (LEMPC) [3]. For example, [5] explored three strategies for attempting to detect attacks on the sensors which involved probing for attacks by modifying the control law over time to drive the closed-loop state toward different operating points in state-space if there is no sensor measurement cyberattack (failure of the state measurements to approach these auxiliary operating points could signify a sensor measurement attack), to compare state predictions with state measurements and flag an attack if these are not sufficiently close to one another, and to compare redundant state estimates with one another and flag an attack if these are not sufficiently close to one another. [4] extended these strategies to the case that the process dynamics change over time. Despite the theoretical guarantees which can be made for these detection strategies in the presence of sensor measurement attacks, they are not theoretically guaranteed to handle attacks on the controller codes, controller-to-actuator communication links, or directly on the process actuators, consequently making any any response to such threats unclear.

Motivated by these considerations, this work seeks to explore how integrated attack detection and control strategies using LEMPC might be developed to handle actuator attacks, as well as simultaneous actuator and sensor attacks. To explore techniques for handling actuator attacks, initially, the utility of the three detection strategies from [5] is explored when the actuators are attacked, with updates to the strategies as necessary. It is demonstrated, for example, that while the probing-based strategy from [5] lacks an ability to conclusively identify attacks on the sensors, a similar strategy is able to ensure closed-loop stability under actuator attacks that are not flagged under sufficient conditions. In contrast, the state estimation strategy from [5] may have difficulty in flagging actuator attacks if the estimates are not impacted by a falsified control action. In addition to extending the attack detection strategies from [5] to the actuator attack case, we also introduce concepts for handling actuator attacks based on checking whether a metric which would signify stability under normal operation (e.g., the Lyapunov function developed around the operating steady-state) is meeting requirements (e.g., decreasing over time), or developing redundant controllers to cross-check control actions received at the actuator. After developing these strategies, we discuss concepts for combining the sensor and actuator attack detection methods, with consideration for the types of guarantees which can be made in this scenario. A chemical process example consisting of a continuous stirred tank reactor is used to illustrate the developments.

References:

[1] Davis, J., Edgar, T., Graybill, R., Korambath, P., Schott, B., Swink, D., Wang, J., Wetzel, J. Smart manufacturing. Annual review of chemical and biomolecular engineering (6), 141-160 (2015).

[2] Lezzi, M., Lazoi, M., Corallo, A. Cybersecurity for industry 4.0 in the current literature: A reference framework. Computers in Industry (103), 97-110 (2018).

[3] Heidarinejad, M., Liu, J. and Christofides, P.D. Economic model predictive control of nonlinear process systems using Lyapunov techniques. AIChE Journal, 58(3), 855-870 (2012).

[4] Rangan, K. K., Oyama, H., & Durand, H. Integrated Cyberattack Detection and Handling for Nonlinear Systems with Evolving Process Dynamics under Lyapunov-based Economic Model Predictive Control. Chemical Engineering Research and Design (2021).

[5] Oyama, H. and H. Durand. Integrated Cyberattack Detection and Resilient Control Strategies Using Lyapunov-Based Economic Model Predictive Control, AIChE Journal (66) ("Futures" issue), ee17084 (2020).

[6] Liu, S., Wei, G., Song, Y., Liu, Y. Extended kalman filtering for stochastic nonlinear systems with randomly occurring cyber attacks. Neurocomputing 207, 708-716 (2016).

[7] Li, Y., Voos, H., Rosich, A., Darouach, M. A stochastic cyber-attack detection scheme for stochastic control systems based on frequency-domain transformation technique, in: International Conference on Network and System Security, Springer, 209-222 (2015).

[8] Huang, K., Zhou, C., Tian, Y.-C., Yang, S., & Qin, Y. Assessing the Physical Impact of Cyberattacks on Industrial Cyber-Physical Systems. IEEE Transactions on Industrial Electronics, 65 (10), 8153–8162 (2018).

630137 Detectability-Based Controller Design Screening for Multiplicative Sensor-Controller Attacks

Abstract:

Increasing reliance on wireless communication and complexity of cyberattacks have rendered industrial control systems vulnerable to cyberattacks by malicious agents [1]. Traditionally, cyberattack threats to industrial control systems have been viewed as an information technology (IT) issue and addressed through IT infrastructure-based approaches. However, over the past decade, several highly sophisticated cyberattacks (e.g., Stuxnet virus (2010), German steel mill attack (2014), Ukrainian power grid attack (2015), TRITON (2017) [2]) have demonstrated that IT infrastructure-based solutions to handling cyberattacks on industrial control systems are insufficient on their own. This realization has led to research on several approaches for handling cyberattacks from a process control perspective. Examples range from designing process control systems (PCSs) that are inherently attack-resilient to developing cyberattack detection, identification and mitigation schemes (e.g., [3]-[6]).

One way an attack may affect a PCS is by manipulating data sent over communication links between the PCS components. While detectability of cyberattacks affecting the PCS communication links may be considered a systems-theoretic property, it is possible for certain controller design parameters to mask the impact of cyberattacks from certain detection schemes. These cyberattacks, called zero-alarm attacks, keep the alarm rate at zero and are particularly challenging to address. However, based on the knowledge of the impact that controller design parameters have on the detectability of an attack, it may be possible to design a controller that does not result in a zero-alarm attack.

Existing control system design criteria typically account for aspects such as closed-loop stability of the process, process economics and robustness to uncertainty [7,8]. However, they do not account for cybersecurity considerations. In this work, we present a novel approach to incorporating the detectability of a cyberattack into the control system design. Specifically, we present a controller screening methodology intended to aid a control designer in identifying and discarding controller and observer design parameters that lead to a multiplicative zero-alarm sensor-controller attack of a certain magnitude on a PCS. We establish the definitions of undetectable (zero-alarm) and potentially detectable attacks for a particular class of residual-based detection schemes. Then, for a given magnitude of multiplicative attack and choice of controller design parameters, we characterize a minimum invariant residual set-based condition that can be used to determine whether the multiplicative attack is undetectable. Leveraging the developed condition, we then present a controller design screening methodology that aids in the identification of controller design parameters leading to undetectable attacks of a certain magnitude. Thus, a controller design that is selected based on the screening methodology ensures that an attack of the chosen magnitude is potentially detectable with respect to the class of residual-based detection schemes considered. Finally, we demonstrate the application of the detectability-based controller screening methodology using a chemical process example.

References:

[1] K. E. Hemsley, E. Fisher *et. al.*, "History of industrial control system cyber incidents", Idaho National Laboratory (INL), Idaho Falls, ID (United States), Tech. Rep. INL/CON-18-44111, 2018.

[2] H. Kayan, M. Nunes, O. Rana, P. Burnap, and C.Perera, “Cybersecurity of industrial cyber-physical systems: A review”, *arXiV:2101.03564.2021.*

[3] S. Chen, Z. Wu and P.D. Christofides, “A cyber-secure control-detector architecture for nonlinear processes”, *AIChE Journal,* vol. 66, no. 5, p. e16907, 2020.

[4] H. Durand and M. Wegener, “Mitigating cyberattack impacts using Lyapunov-based economic model predictive control”, in *Proceedings of the American Control Conference,* Virtual, 1-3 July 2020, pp. 1894-1899.

[5] J. Giraldo, D. Urbina, A. Cárdenas, J. Valente, M. Faisal, J. Ruths, N. O. Tippenhauer, H. Sandberg and R. Candell, “A survey of physics-based attack detection in cyber-physical systems”, *ACM Computer Surveys,* vol. 51, no. 4, pp. 76:1 – 76:36, 2018.

[6] N. Hashemi and J. Ruths, “Co-design for security and performance: LMI tools”, *arXiv:1909.12452,* 2019.

[7] P.D. Christofides, N. H. El-Farra, “Control of nonlinear and hybrid process systems: Designs for uncertainty, constraints and time-delays”, Vol. 324, Springer Science & Business Media, 2005.

[8] J.A. Romagnoli, A. Palazoglu, “Introduction to process control”, CRC press, 2020.

632836 Expanding Resilient Lyapunov-Based Economic Model Predictive Control Concepts to a Distributed Control Framework

Abstract:

Distributed control is a technique with potential advantages in terms of computation time and tractability in controlling large-scale cyber physical systems. From a cybersecurity perspective, however, the increased number of information-passing connections relative to centralized control increases the opportunity for attackers to gain access to a component of a distributed system, which may impact process safety and/or profitability [2]. Specifically, false state measurements and other types of attacks have the potential to impact the performance of one or many controllers in the network. At the same time, distributed control offers more controllers with the potential capability to cross-check one another, ideally to pinpoint ones that cause problems on the network and remove them. To be able to analyze resilience of distributed controllers to attacks, it is desirable to utilize a distributed control framework that has strong closed-loop stability and feasibility guarantees in a centralized context. Lyapunov-based economic model predictive control (LEMPC) [5] is an optimization-based control technique that is able to guarantee closed-loop stability and recursive feasibility even in the presence of sufficiently small bounded disturbances. Already, several works have looked at how LEMPC relates to control system cybersecurity. For example, centralized LEMPC is explored with respect to guarantees which can be made in the presence of attacks for centralized control designs in [3] and [4]. Machine learning and data based methods related to cybersecurity have also been considered in conjunction with LEMPC [1]. However, despite a strong theoretical basis for distributed LEMPC [6, 7], integrating this framework with the detection policies for sensor measurement cyberattacks in [4] has not yet been explored, and further investigation of how the distributed framework might lend itself to means for isolating problematic controllers on the network with stability guarantees also must be performed.

Motivated by these considerations, this talk will extend three cyberattack detection strategies of [4] to a distributed control framework (both iterative and sequential LEMPC [6, 7]). The stability guarantees will be characterized, and in particular the challenges with diagnosing which of the distributed controllers is experiencing an attack if there is a discrepancy compared to what is expected (if, for example, the Lyapunov function should decrease in each of a set of sequential distributed controllers, and is found not to in one in the sequence). We also discuss a methodology, inspired by [8], for removing a controller that might have been attacked (e.g., re-coded) from a network (as in [8], it may be possible to operate the process with a reduced set of control actions if returned to a region around the steady state), and then seeking to introduce a level of randomization in the manner in which control actions are subsequently computed between the isolated controller and the remaining controllers to attempt to make it harder for an attacker to remove specific controllers in an attempt to generate a characterizable behavior with the network.

[1] Chen, Scarlett, Zhe Wu, and Panagiotis D. Christofides. "Cyber-attack detection and resilient operation of nonlinear processes under economic model predictive control." *Computers & Chemical Engineering* 136 (2020): 106806.

[2] Chen, Scarlett, Zhe Wu, and Panagiotis D. Christofides. "Cyber-security of centralized, decentralized, and distributed control-detector architectures for nonlinear processes." *Chemical Engineering Research and Design* 165 (2021): 25-39.

[3] Durand, Helen. "A nonlinear systems framework for cyberattack prevention for chemical process control systems." *Mathematics* 6.9 (2018): 169.

[4] Oyama, Henrique, and Helen Durand. "Integrated cyberattack detection and resilient control strategies using Lyapunov-based economic model predictive control." *AIChE Journal* 66.12 (2020): e17084.

[5] Heidarinejad, Mohsen, Jinfeng Liu, and Panagiotis D. Christofides. "Economic model predictive control of nonlinear process systems using Lyapunov techniques." AIChE Journal 58.3 (2012): 855-870.

[6] Liu, Jinfeng, Xianzhong Chen, David Muñoz de la Peña, and Panagiotis D. Christofides. "Sequential and iterative architectures for distributed model predictive control of nonlinear process systems." AIChE Journal 56, no. 8 (2010): 2137-2149.

[7] Albalawi, Fahad, Helen Durand, and Panagiotis D. Christofides. "Distributed economic model predictive control for operational safety of nonlinear processes." AIChE Journal 63.8 (2017): 3404-3418.

[8] Lao, Liangfeng, Matthew Ellis, and Panagiotis D. Christofides. "Smart manufacturing: Handling preventive actuator maintenance and economics using model predictive control." AIChE Journal 60.6 (2014): 2179-2196.

635087 Commercialization of Aptamer-Based Electrochemical Sensors

Abstract:

QSM Diagnostics is in the process of commercializing electrochemical sensor technology developed by the Goluch Group at Northeastern University. The initial application of the technology is for rapidly diagnosing and monitoring bacterial infections in companion animals. This is accomplished by detecting quorum sensing molecules (QSMs) that are present in specimens collected from the animals. The platform technology, however, can be adapted to almost any target molecule of interest by simply exchanging the recognition element that is attached to the sensor. The QSM sensing platform uses aptamers as recognition elements in conjunction with square wave voltammetry. Theoretically, any properly developed aptamer can be integrated to create a custom rapid, quantitative, testing system.

As an example of this turn-key solution capability, we have now demonstrated the detection of moxifloxacin, a broad-spectrum fluoroquinolone antibiotic, in human serum samples. This presentation will describe the process by which the new sensor was developed and tested as well as the challenges associated with the development process.

635137 (Invited Talk) Developing Nanosensor Technology to Detect Cancer in Patients

Abstract:

We have been developing and translating a technology to detect biomarkers quantitatively in humans to enable the early diagnosis of ovarian cancer. There is no effective screening test for ovarian cancer and no reliable method of early detection. Approximately 85% of patients with ovarian cancer will have widely disseminated peritoneal disease at presentation, resulting in five-year survival rates of < 30%. This is in stark contrast to patients diagnosed with early-stage localized disease, where survival rates are over 90%. The early detection of ovarian cancer would markedly increase survival rates. Serum biomarkers have repeatedly failed to identify early disease or change mortality rates, in part due to low specificity. We believe that the measurement of biomarkers using an implantable device in the uterine cavity will allow for discrimination of benign states from malignant disease. We are working in collaboration with clinicians to develop an implantable sensor platform for protein biomarkers capable of minimally-invasive, highly-sensitive, reproducible, in vivo measurements in disease models. We demonstrated a device to implant the sensors in animals and resected human uteri. Our goal is to develop the capability for real-time and progressive patient-centered monitoring of biomarkers for early ovarian cancer detection in at-risk patients.

630605 (Invited Talk) Wearable Health Sensors

Abstract:

The need for well-managed hydration and the monitoring of physiological parameters in real-time is not only crucial for athletes to optimize performance but to also prevent illness and injury from occurring. Understanding sweat loss and calculating hydration requirements to avoid dehydration or overhydration is a significant challenge for athletes. This is because sweat losses vary significantly from athlete to athlete both in terms of concentration of electrolytes and total volume of sweat lost. When sweating, an individual may lose as little as ~200mg of sodium per liter or as much as ~2000 mg. This variance, combined with significant differences in sweat volume, lead to vastly different hydration requirements among individuals. The RooSense Optimal Sweat Performance (OSP) Band is the first lightweight fabric sensor to provide real-time information regarding hydration levels during exercise or training through selective determination of sodium levels, a better marker for athletic performance, safety, and injury.

Monitoring sweat is a compelling choice to gain insight into a person's health at the molecular level, as sweat is rich in physiological and metabolic information that can be obtained non-invasively. In spite of advances made towards the detection of biomarkers in sweat, there is no sensor capable of long-term detection in constricted or load-bearing applications where other flexible plastic sensors might cause discomfort. Textile-based and nanocomposite technologies, such as those developed at RooSense LLC, provide an optimum scaffold for monitoring biomarkers at the surface of the skin as they do not influence the sweat composition or sweat rate and can be seamlessly integrated for wearable sensing and translational medicine applications.

This presentation will focus on the development of a prototype wearable hydration monitor, validation of the prototype in the laboratory, on body testing in an exercise physiology laboratory.

634910 (Invited Plenary Talk) Silicon Nanowire Photoelectric Protein Sensors

Abstract:

Protein sensing is where the rubber hits the road in biology. Hence measuring protein concentrations is important for many applications including point of care testing, bio and pharma manufacturing, food production, bioweapons detection, and fertility assistance. The field of biosensing has made great progress and is about to enter a more mature stage that will greatly improve our lives.

In this talk, we review some of the types of biosensing being developed. Then we focus on our photo-electric ELISA. Our photo-electric ELISA uses arrays of millions of vertically aligned silicon nanowires to detect dilute qualities of protein in a solution. Protein measurements can be completed in less than 15 minutes and the sensors can be made a very low cost. The high surface area to volume ratio allows for a sensor that can detect dilute quantities of protein in a solution. Since we use silicon wafers and perform an electrical measurement, we can make use of the semiconductor's knowledge of packing many tests onto the same chip and thus can form a multiplexed test that measures the concentration of many different proteins on the same chip. Since we do an electrical measurement, our measurement system is low cost and can operate for a wide range of vales.

We will explore the fabrication of the nanowire sensors and the mechanism of operation of the sensors. In addition, we will present results showing the response of the sensor when exposed to the protein of interest. We will then discuss how we are optimizing the sensor and improving the reproducibility of the test.

635047 (Invited Plenary Talk) Glucose Sensors for Bioprocessing: Progress and Perspectives

Abstract:

The ability to continuously monitor glucose concentrations in the same manner as pH, temperature, and dissolved oxygen has been a goal for bioprocessing for decades. The importance of continuous glucose monitoring has grown in recent years with the impetus of

FDA's Process Analytical Technology initiative, advances in continuous cultivation technologies, and findings that optimal productivity and product quality require tight control over substrate concentrations. A variety of glucose monitoring strategies have been proposed and some commercialized, ranging from glucose-specific sensors to spectroscopic analyzers and from in-situ measurements to at-line systems with automated sampling. This presentation will review the history of glucose sensing for bioprocessing, current options, and some perspectives on the future.

634817 (Invited Plenary Talk) Optical Sensing and Cell Culture

Abstract:

Real-time monitoring of pH, cell density, and dissolved oxygen concentration with optical sensors is challenging traditional biological cell culture assumptions. By providing insights into cellular metabolism and localized environmental conditions, optical sensing of critical process parameters has already proved useful for the analysis, experimental design, and manufacturing control in a variety of fields, simultaneously improving results, increasing reproducibility, and decreasing cost. Minimally invasive and non-invasive sensing modalities are changing long-held cell culture paradigms and results are being visualized in real-time. FDA's Process Analytical Technology (PAT) initiative outlined in a guidance document issued in 2004 encourages manufacturers to use the latest scientific advances in pharmaceutical manufacturing and technology and implement continuous real time quality assurance. The use of PAT tools such as process analyzers that measure critical quality attributes and process control tools is encouraged and ideally these tools should be introduced during the development phase. Monitoring, understanding, and regulating critical cell culture parameters has the potential to make scale-up easier, faster, and less costly, easing the path from laboratory research to realized industrial application and decreasing time from bench to bedside.

631529 Multiscale Modeling of Spray Coating of Perovskite QDs: Understanding the Role of Molecular Interactions in Particle Aggregation.

Abstract:

The US Department of Energy has set high-priority targets to reduce the cost of photovoltaics from ~ 10 cents/kWh to half of that at 5 cents/kWh by 2030. This has fueled a plethora of research in the discovery, optimization, and commercial production of high-efficiency solar cells. Perovskite solar cells (PSCs), especially $CsPbBr_3$, have received special attention due to their high quantum efficiency and tunable absorption/emission spectra [1]. Often, the perovskite quantum dots (QDs) are synthesized using one of the various batch synthesis routes, and after appropriate post-processing, they are spin-coated and separately cured to make a functional layer for the PSCs [2]. However, the problems of batch-to-batch variation, low-material utilization, and separate curing required in the spin coating process have slowed down and made the commercial production of PSCs difficult. In response, recent studies have demonstrated that spray coating can counter these challenges associated with spin coating by (a) providing a fast-scalable continuous production alternative with better material utilization; (b) offering a large control space for uniform deposition without the need for a separate curing step [3], and (c) integrating it with a continuous crystallizer to result in higher production capacity [4]. That being said, modeling and control of spray coating is a non-trivial task due to a combination of macroscopic phenomena (e.g., transport and thermal), and molecular-level interactions (e.g., dipole-dipole, van der Waals, and steric forces), thereby requiring a multiscale modeling approach.

In spray coating, the colloidal solution from the continuous crystallizer is atomized and sprayed onto a thermally heated silicon substrate through a pressurized nozzle. Consequently, the atomized droplets impinge onto the substrate and form a liquid splat followed by evaporation of the solvent (toluene). In this work, the spray coater is modeled using a system of macroscopic mass and energy balance equations (MEBE), which describe the effect of surface tension, vapor pressure, nozzle velocity, and the size of the impinging droplet on solvent evaporation. Further, during the solvent evaporation, the QD particles undergo particle aggregation due to the presence of various molecular interactions, and can lead to the formation of a non-uniform film, and demonstrate the coffee-ring effect. Thus, a modified form of the discrete element method (DEM), which solves newton's equation of motion for each QD particle, is utilized to accurately model these microscopic interactions [5]. Specifically, the DEM considers four molecular interactions, viz., steric hindrance, dipole-dipole forces, Brownian motion, and van der Waals forces. This is because, (a) the QD particles have a long-chain ligand-shell around them, which results in steric hindrance (repulsion) if two QD particles come very close to each other during aggregation; (b) experimental observations show a non-zero dipole moment for the QD particles [6]; and (c) the QD particles have a size range of 5-10 nm, which makes them very susceptible to undergo Brownian motion. Finally, the continuum model (MEBE) and the DEM are dynamically coupled to develop a high-fidelity multiscale model of spray coating.

The simulation results generated for different nozzle velocities and QD concentrations demonstrate the typical coffee ring formation as seen in the experiments [7]. The results also elucidate the effect of controllable macroscopic variables (i.e., substrate temperature, nozzle velocity, and QD particle concentration) on microscopic film evolution (i.e., aggregate formation, fractal structure and size). Overall, the proposed multiscale modeling framework provides a mechanistic understanding of the role of molecular interactions in QD particle aggregation and film deposition during temperature-controlled spray coating of QD particles for application in PSCs. Furthermore, this work provides a general platform for multiscale modeling of spray coating of different particle systems and will be incorporated with advanced optimization or process control frameworks to ensure adequate set-point (film thickness and smoothness) tracking in future works.

Literature Cited:

1. Protesescu L, Yakunin S, Bodnarchuk MI, Krieg F, Caputo R, Hendon CH, Yang RX, Walsh A, Kovalenko MV. Nanocrystals of cesium lead halide perovskites ($CsPbX_3$, X= Cl, Br, and I): novel optoelectronic materials showing bright emission with a wide color gamut. Nano Letters. 2015 Jun 10; 15(6): 3692-6.
2. Chen LC, Tien CH, Tseng ZL, Ruan JH. Enhanced efficiency of $MAPbI_3$ perovskite solar cells with FAPbX3 perovskite quantum dots. Nanomaterials. 2019 Jan; 9(1): 121.
3. Yuan J, Bi C, Wang S, Guo R, Shen T, Zhang L, Tian J. Spray-Coated Colloidal Perovskite Quantum Dot Films for Highly Efficient Solar Cells. Advanced Functional Materials. 2019 Dec; 29(49): 1906615.
4. Sitapure N, Epps R, Abolhasani M, Kwon JS. Multiscale Modeling and Optimal Operation of Millifluidic Synthesis of Perovskite Quantum Dots: Towards Size-Controlled Continuous Manufacturing. Chemical Engineering Journal. 2020 Dec 7: 127905.
5. Peng Z, Doroodchi E, Evans G. DEM simulation of aggregation of suspended nanoparticles. Powder Technology. 2010 Dec 10; 204(1): 91-102.
6. Wang W, Zhang Y, Wu W, Liu X, Ma X, Qian G, Fan J. Quantitative modeling of self-assembly growth of luminescent colloidal $CH_3NH_3PbBr_3$ The Journal of Physical Chemistry C. 2019 Apr 28;123(20):13110-21.
7. Xu T, Lam ML, Chen TH. Discrete Element Model for Suppression of Coffee-Ring Effect. Scientific Reports. 2017 Feb 20;7(1):1-0.

624233 Findings and Conclusions from a Mobile Worker and Augmented Reality Enabled Continuous Manufacturing Skid Project.

Abstract:

This abstract presents the final conclusions regarding a project initially presented at AIChE 2021 Spring Meeting as an oral presentation (abstract#618949) and poster (abstract#623026).

Background:

Manufacturing USA was created in 2014 to secure U.S. global leadership in advanced manufacturing by connecting people, ideas, and technology. Manufacturing USA institutes convene business competitors, academic institutions, and other stakeholders to test applications of new technology, create new products, reduce cost and risk, and enable the manufacturing workforce with the skills of the future (www.manufacturingusa.com).

MxD encourages factories across the United States to deploy digital manufacturing and design technologies, so those factories can become more efficient and cost-competitive (www.mxdusa.org).

Project detail:

This abstract relates to a continuous process manufacturing skid that has a focus on mobile workers and demonstrates the future of manufacturing by using a digital twin enabled physical skid build. The question the skid answers is "How would you visualize sensor data with the best User Interface/User Experience design to get a seamless experience in continuous manufacturing?"

The findings and conclusions come from a project team consisting of an government owned site personnel, a private chemical processing company, a multi-national technology supplier and multiple service providers. Equipment and assets applied in the project go from sensors to process equipment to pumps/drives to valves to a DCS to a MES and complete engineering and design suite and finally to an IT/OT-integrated and IIoT-enabled setup to support day-to-day operations and mobile workers. Cloud platform with Artificial Intelligence / Machine Learning capabilities is connected to the site and the skid.

Currently there is considerable work done in chemical industry to study the current state and possible future states when it comes to skids, processes, plants, companies, professionals, workers and value chains. Applying novel techniques into operations is in practice an efficient way for human beings to get a real grasp on what is impacted, how it is impacted and what do we want to do with the novel capabilities that come from innovation areas such as the wider digital enterprise or industry 4.0 concept. Concrete studies like this one bring important contributions to the field, especially when introduced as an example in a discussion or presentation at an event such as the AIChE Annual Meeting.

The subject of this abstract was originally planned and divulged in 2019, moved into a contractual obligation in 2020 and is scheduled to conclude in summer of 2021. This timeline allows the authors to go through the whole lifecycle from problem statement to selected methods to real physical results and finally to discuss and expand the implications for future ways of working.

The audience will learn about augmented reality and digital twin use cases around a continuous process testbed. The audience will understand that this skid is available for industry professionals and students to visit at a public facility on Goose Island, Chicago.

Use Cases:

The goal of the process industry test bed skid (physical) and its digital twin (virtual) is to enable field operators, engineers, factory managers, and other key manufacturing roles to gain data-driven insights for increased operational awareness in order to better solve manufacturing problems and optimize processes. A factory manager should be able to communicate with and assign tasks to the mobile worker and view all data necessary to make informed decisions about the assignment of tasks. The solution will be demonstrated using the process test bed and should be agnostic to sensor type, analytics solution, and control type. Visualization of this data should include, but is not limited to, dashboards for factory managers and augmented reality technology for field operation using mobile devices (e.g. tablet and smart glasses).

Conclusions: The three authors of the abstract will each share their findings from their viewpoint, those being the viewpoints of an owner/operator company, the production site and the main technology provider.

632817 Practical Issues in Cybersecurity: From Encryption to Images

Abstract:

Cybersecurity has gained interest for chemical processes in recent years. For example, [1] discusses techniques for considering cybersecurity during traditional chemical engineering procedures such as hazard and risk analysis. One of the main drivers for the cybersecurity discussion is an increased interest in recent years in concepts such as Industry 4.0 [2] and the Industrial Internet of Things [3], and a general increase in data analytics and artificial intelligence approaches. Future work in this domain would benefit from research that enables a more integrated approach to control system cybersecurity in the chemical engineering research field, in which the cyber defenses at a plant are more explicitly considered, as well as advanced control frameworks and concepts. These concepts include resilient frameworks, to analyze how a system functions as a whole and the types of attacks it remains vulnerable to after both, the theoretical requirements for certain resilient control frameworks as well as, practical communication and networking protocols that need to be accounted for. In addition, there are a wide variety of issues becoming more important to consider from a cybersecurity perspective over time. For example, image analysis is important for certain types of decision-making systems, though image data represents another type of information that could be falsified to manipulate such decision-making [4], and computations on the Cloud, where it may be desired to encrypt not only the data but the computations themselves [5, 6], are additional concerns.

Motivated by the above, this talk highlights multiple concepts in the practical aspects of control system cybersecurity as it interfaces with advanced control. For example, using a case study of a benzene hydealkylation process under distributed Lyapunov-based economic model predictive control [7], we explain the role of encryption protocols in enabling safe transfer of information between distributed units, and the extent to which the encryption might enable attacked controllers to be identified on the network. We utilize a process example of a semi-batch polymerization reactor under model predictive control with uncertainty scenarios [8] and offset-free model predictive control [9] to demonstrate the manner in which disturbance-handling is not equivalent to cyberattack resilience, and we also discuss profit targeting as a key metric in distinguishing attacks and faults. We follow up with a discussion of considerations in image analysis, which form part of the increasingly automated world that is more open to attacks, and verification in that domain, and how such concepts relate to cybersecurity and cyberattack resilience. For example, focusing on the zinc flotation process, in which image-based control is considered today [10], we explore the potential use of software such as Blender for creating a preliminary closed-loop simulation methodology for simulating the zinc flotation process, including both system state evolution under control actions, as well as image generation and processing, along with analysis on the impacts of cyberattacks on image data that determine process characteristics. Finally, we will discuss the role of communication techniques, as well as encryption techniques which facilitate communication with the Cloud, in control for nonlinear chemical process systems.

[1] Cormier, A., & Ng, C. (2020). Integrating cybersecurity in hazard and risk analyses. *Journal of Loss Prevention in the Process Industries*, *64*, 104044.

[2] Reis, M. S., & Kenett, R. (2018). Assessing the value of information of data-centric activities in the chemical processing industry 4.0. *AIChE Journal*, *64*(11), 3868-3881.

[3] Sisinni, E., Saifullah, A., Han, S., Jennehag, U., & Gidlund, M. (2018). Industrial internet of things: Challenges, opportunities, and directions. *IEEE Transactions on Industrial Informatics, 14*(11), 4724-4734.

[4] Bucci, E. M. (2018). Automatic detection of image manipulations in the biomedical literature. *Cell death & disease, 9*(3), 1-9.

[5] Darup, M. S., Redder, A., & Quevedo, D. E. (2018). Encrypted cloud-based MPC for linear systems with input constraints. *IFAC-PapersOnLine, 51*(20), 535-542.

[6] Alexandru, A. B., Morari, M., & Pappas, G. J. (2018, December). Cloud-based MPC with encrypted data. In *2018 IEEE Conference on Decision and Control (CDC)* (pp. 5014-5019). IEEE.

[7] Liu, J., Chen, X., Muñoz de la Peña, D., & Christofides, P. D. (2010). Sequential and iterative architectures for distributed model predictive control of nonlinear process systems. *AIChE Journal, 56*(8), 2137-2149.

[8] Lucia, S., Finkler, T., & Engell, S. (2013). Multi-stage nonlinear model predictive control applied to a semi-batch polymerization reactor under uncertainty. *Journal of process control, 23*(9), 1306-1319.

[9] Das, B., & Mhaskar, P. (2015). Lyapunov-based offset-free model predictive control of nonlinear process systems. *The Canadian Journal of Chemical Engineering, 93*(3), 471-478.

[10] Kaartinen, J., Hätönen, J., Hyötyniemi, H., & Miettunen, J. (2006). Machine-vision-based control of zinc flotation—a case study. *Control Engineering Practice, 14*(12), 1455-1466.

634112 Keynote Talk: Machine Learning and AI Applications in the Chemical Industry

Abstract:

Industrial chemical processes usually come with abundant data and thus provide huge opportunity to leverage machine learning and AI techniques for process improvement. Over the past decade as computational resources have become readily available at much lower costs, together with rapidly growing availability of cloud services, sensor and analytical technology, the number of AI applications adopted for process monitoring, control and optimization have also increased exponentially. Aside from traditional multivariate analysis that is often used for monitoring and fault diagnosis, other analytics fields such as (but not limited to) image analysis, deep learning, sensor fusion, text mining, reinforcement learning, hybrid and physics informed surrogate modeling are also being actively applied to increase reliability, reduce safety events, predict the onset of potential equipment failures and enable real time optimal decision making. This talk will provide examples of AI applications using industrial case studies and also address current challenges as well as outlook.

634113 Keynote Talk: Deploying AI for Automated Monitoring of Physical Infrastructure

Abstract:

This presentation will discuss how AI can be used for automated monitoring of physical infrastructure across various assets and processes. Key topics include:

The purpose and benefits of following an Op-Specific approach to digital transformation based on survey results.

How an operational data platform can reduce resource consumption, improve sustainability goals, and increase revenue.

Steps to implement a specific use-case.

An Operational Data Platform (ODP) can connect the constantly changing realities of the physical world with the power of intelligent software. Plutoshift has built an ODP to unlock the potential of our operational data to enable process monitoring and remote decision making. Plutoshift serves a variety of industries such as chemicals, food & beverage, manufacturing, water treatment, and more. The Plutoshift platform has enabled significant cost reductions ranging from 20-30%. Along with that, it has enabled millions of dollars of revenue increases for customers. The platform is built to help companies achieve sustainability targets as well. For example, the platform helped one global brewer's facility save close to 16 million gallons of water in a single year. These savings are being driven by enhanced visibility of assets and processes across the physical infrastructure.

630099 High Strength, High Toughness Parts Via Dual Material Fused Filament Fabrication

Abstract:

Previous research in the field of bicomponent thermoplastic 3D printing material showed promise in developing structures with tough, near-isotropic properties; a feat that is unheard of in tradition fused filament fabrication (FFF) systems. The combination of a core with a significantly higher glass transition temperature (T_g) than the sheath allows for parts that can be annealed to increase chain diffusion across layers while maintaining dimensional stability. The use of a thermal draw tower in previous research allowed for rapid prototyping of material and geometric combinations, but lacks viability as a larger scale manufacturing process due to the bottleneck of creating dual material preforms and requiring secondary processing. Using a convention coextrusion technique, with an additive manufacturing driven iteration process, showed promise as a scale-up technique, and was able to produce larger quantities of consistent filament needed for more comprehensive analysis and production. Compared to current state of the art ABS filaments, this filament showed a 5 time increase in z- direction impact toughness, and a 3 time increase in z-direction tensile strength after annealing. The enhanced thermal stability of this material allowed for higher print temperatures which also significantly improved strength. This research continues to demonstrate the groundbreaking improvements to FFF that are possible through dual material filaments as well as the manufacturing feasibility of such a product.

630310 Process-Structure-Property Relationships in Additively Manufactured Polypropylene Blends

Abstract:

Material extrusion (MatEx) based additive manufacturing (AM) of polymers has gained great popularity due to its versatility, operational simplicity, and cost-effectiveness. However, AM or 3D printing of semicrystalline polymers are still largely relegated to prototyping applications. Major problems involving shrinkage and warpage of the printed parts must be addressed before such polymers can be used for printing functional parts that can be employed in load bearing applications. Moreover, there can be a great deal of variation in the morphology and crystallization behavior of the polymer depending upon the processing conditions employed and the inherent material properties. Thus, the goal of this work is to provide fundamental understanding of the complex changes that take place during MatEx-based printing of semicrystalline polymer blends and thereby assist in developing tailored compositions for consistent and reliable AM processing.

In the present work, we blend polypropylene (PP) with polyolefin-based copolymers and evaluate the compatibility of the blends with MatEx-based AM. Differences in crystallization halftimes and crystallographic structures have a profound impact on the interlayer adhesion and residual stress state, which directly controls the mechanical performance and warpage of the printed parts. We investigate the effect of high shear in the printer nozzle on the flow-induced crystallization of the blends and couple with the melt flow behavior through detailed rheological characterizations. Variations in the crystal structure, degree of orientation, and degree of crystallinity, resulting from different printing conditions, significantly influence the tensile properties of the printed parts. The findings from this study can be leveraged in toolpath planning, process parameter optimization, and new feedstock development for MatEx, highlighting the current limitations as well as providing valuable insights into necessary processing modifications in order to enable MatEx of next generation semicrystalline polymers.

629126 Understanding Flow and Stress Development in 3D Printing By Material Extrusion

Abstract:

Additive manufacturing of plastics offers a plethora of potential advantages associated with primarily mass customization, designs incompatible with injection molding, and distributive manufacturing. 3D printing of thermoplastics by Material Extrusion (MatEx) using filaments is well established, but the properties of the printed parts tend to be inferior to those of parts produced by conventional plastic manufacturing techniques. Defects (voids) and poor interface development between printed roads tend to be blamed for the poor performance. Additionally, the transient nature of the print leads to highly non-isothermal processing histories and potential chain alignment as the cooling rates are rapid and can be similar to relaxation times for the polymers. Moreover, the print path involves rapid acceleration and deceleration of the print head around corners, which will influence the viscoelastic properties of the polymer as it is printed. Here we directly visualize the flow in simple printed coupons using a plug of pigment in the filaments to examine the role of layer height (compression) and corners in the printed object. These results illustrate the formation of waves in the apparent streamlines from the flow indicator near the edges of even simple sample. These effects can extend over longer lengths than expected based on the simulated print path and suggest that elastic effects in the melt can be significant. The inclusion of transition metals in the pigments provides x-ray contrast to enable the visualization of the flow profiles with x-ray tomography in addition to optical observations, which provides 3D rendering of the flow profiles around the corners of the print. The complex temperature history will also influence the dimensional accuracy of the printed part through the stress development during solidification. We will briefly describe how the printed shape influences these effects and the reported mechanical properties.

631518 Autonomic Self-Healing of 3D Printed Polymer Composites

Abstract:

One approach to a more sustainable materials ecosystem is through extending material lifetimes through self-healing behavior. While self-healing has utility across many applications, it is particularly important for components fabricated for hard to reach or hard to replace applications. Concurrently, the potential of additive manufacturing has grown tremendously in recent years and the benchmarking and exploration of new polymer materials and polymer composite materials for 3D printing applications is critical. Notedly, many 3D printed specimens tend to show inferior mechanical strength, and thereby increased potential for material damage and failure during fabrication and use, compared to traditional processing approaches. The inclusion of self-healing behavior is a promising approach to address the microcracks that can develop within the structure where fracture detection is difficult, and repair is almost impossible both during fabrication and in use. Moreover, many of the polymer material options for 3D printing are not intrinsically recyclable, requiring disposal if they suffer any damage or degradation, increasing environmental waste through the premature end of their useful life. Here, we incorporate self-healing properties towards extending the lifetimes of 3D printed polymeric objects. Inspired by biological self-healing, where a damage event triggers an autonomic healing response, microcapsules containing healing fluids are embedded within the 3D printed composites. Microcapsules rupture during a damage event release the healing agent, and heal the surrounding material by polymerization, entanglement, or cross-linking without external intervention increasing product life and improving material sustainability and long-term cost. Double shell wall polyurethane/poly (urea-formaldehyde) microcapsules are synthesized by in-situ-interfacial polymerization. Microcapsules with both solvent and monomer core fluids are prepared to investigate both solvent-healing and monomer self-healing mechanisms. Microcapsules, of varied size and at varied concentration, containing healing agents are either incorporated into the host polymer matrix (SLA 3D printing) or are coated onto 3D printing polymer filaments (FDM 3D printing). Microcapsule distribution within composites is visualized using X-ray Nano-CT imaging. Printed composite self-healing behavior is evaluated by their fracture toughness healing efficiency. Overall, we find the inclusion of autonomic self-healing behavior through this approach to be successful for both SLA and FDM 3D printing.

631028 Synthesis, Characterization, and Application of Novel Surface-Eroding Photopolymer Formulations

Abstract:

Three-dimensional (3D) printing has gained popularity in recent years for its ability to produce personalized 3D parts on demand with minimal supplies and with specific functionalities. Photocurable resins, typically consisting of a combination of multi-functional monomers and oligomers, a photoinitiator, and an optical absorber, are formulated for numerous applications in light-based 3D printing. Several groups have studied the application of water-soluble resin formulations, with a common concern being the swelling of the polymer matrix during the dissolution process [1]. Previous efforts to incorporate surface-eroding oligomers into resin formulations has been further limited by the commercial availability of surface-eroding constituents that are suitable for use in light-based 3D printing.

In this work, a synthesis protocol was invented for the production of unique surface-eroding oligomers that later react to form crosslinked networks during the 3D-printing process and readily degrade in the presence of water. Addition of particular constituents in the reaction mechanism at different stages influences the physiochemical behavior of the oligomer in unique environments by incorporating different degrees of hydrophobicity into the oligomer backbone. Novel surface-eroding oligomers are available as both a liquid and solid at room temperature as a result of the devised synthesis protocol. Further incorporation of the surface-eroding oligomers into novel resin formulations provides for high-resolution 3D-printable methacrylic-anhydride-based resins that chemically degrade in the presence of water. Chemical degradation leads to surface-erosion of the crosslinked network, which provides ample opportunity for a wide-array of applications where swelling of the polymer matrix has previously been a significant drawback in the functionality of water-susceptible 3D-printed parts. A commercially available LCD printer was modified to substantially reduce the volume of resin that is required for printing, minimizing costs of testing resin formulations.

References

[1] Robert Liska, Florian Schwager, Christine Maier, Rafael Cano-Vives, and Ju¨rgen Stampfl. Water-soluble pho- topolymers for rapid prototyping of cellular materials. Journal of applied polymer science, 97(6):2286–2298, 2005.

Copyright © American Institute of Chemical Engineers. All rights reserved.

631982 Designing New Printable Thermoset Shape Memory Polymers Via Molecular Simulation and Machine Learning

Abstract:

Printable thermoset shape memory polymers (TSMPs) offer the opportunity to substantially expand the range of applications available to thermoset polymers. A TSMP, after a thermomechanical cycle (i.e., programming), can recover its original shape by heating. The act of programming reduces conformational entropy in the network and increases enthalpy through various interatomic interactions, thus supplying the thermodynamic driving force for shape recovery. The classic approach for developing new polymers is one of trial-and-error, but this brute-force strategy is costly when searching in a large compositional space. By fingerprinting TSMPs at multiple length scales and investigating a significant initial dataset using molecular simulation, we have identified salient features in epoxy/amine-based thermoset shape memory polymers that produce a large shape memory effect, identified storage mechanisms for entropy and energy in TSMPs, and proposed new molecules that we predict will result in improved recovery stress and recovery ratio.

634114 Keynote Talk: Online Measurements for the Petrochemical Industry: Industry 4.0 Trends and Unmet Needs

Abstract:

The primary objectives of the petrochemical manufacturing industry are to operate plants safely, profitably, and reliably. To achieve these objectives, several constraints need to be managed – such as raw material yield, manufacturing costs, and utility requirements. Online measurements via instruments that measure the physical properties and analyzers that measure the chemical properties of the process are critical to the optimized management of these constraints.

Globally, Dow owns and supports > 1 million instruments and > 20,000 online analyzers. Data from these measurements are indispensable to the operation and control of our plants. In this talk we will discuss some of our key requirements for both classes of online measurements, share examples of meaningful advances afforded by Industry 4.0, and articulate key challenges that remain – a call for continued innovation.

628357 Woodchip Moisture Content Estimation Using Short-Range Iot Wi-Fi for the Pulp & Paper Industry

Abstract:

The pulp and paper industry is the third-largest consumer of energy in the US industrial sector thereby leading to tremendous opportunities to improve its energy efficiency and productivity. The pulping process, which converts woodchips into pulp by displacing lignin from cellulose fibers, is one of the most important operations in a pulp and paper mill. For the pulping process, it is important to have real-time knowledge about the moisture content (MC) of the woodchips so that the process can be optimized and/or controlled correspondingly to achieve satisfactory product quality while minimizing the consumption of energy and chemicals. There have been attempts to develop real-time woodchip MC sensors, such as capacitance, resistance, microwave and NIR-based sensors. But these techniques have not been adopted by the industry due to unsatisfactory performance and/or high maintenance requirements that can hardly be met in a manufacturing environment. Also, some of the accurate methods are often lab-based that cannot be implemented online because of time constraint [1]. Due to these limitations, at present, the moisture level is measured only a few times throughout the year in the pulp and paper industry.

To address these limitations, we propose a non-destructive, economic, and robust approach using industrial Internet-of-Things (IIoT) based Wi-Fi and employ channel state information (CSI) to estimate MC in woodchips. We extract CSI by modifying the open-source device drivers for Intel Wi-Fi link 5300 Network Interface Card based on CSITool [2]. CSI contains information about the channel in the form of individual data sub-carriers capturing indoor channel features such as the effect of scattering, fading, and power decay with distance. We hypothesize that when a Wi-Fi signal passes through wood chips, the water or MC in woodchips would cause a corresponding change in the Wi-Fi signal which is then captured in CSI. The goal is to develop a data-driven model to extract the relationship between the change in CSI and the woodchip MC. CSI has been successfully used for moisture classification in wheat [3]. However, compared to wheat moisture classification, woodchip moisture estimation is much more challenging due to the significant heterogeneity in the shape, structure, and arrangement of the woodchips. For example, the woodchip arrangement in the container is expected to have a significant impact on the CSI data. In addition, CSI data have only been used in classifications, which is much easier than regression modeling.

In this work, we propose a regression-based approach to estimate the moisture content in wood chips. To the best of our knowledge, this is the first work to address the continuous estimation of MC using CSI data. Our goal is to develop a robust model that is insensitive to the heterogeneity of the woodchips, particularly the confounding factors such as woodchip shape and the effect of shuffling, to achieve a consistent and accurate MC estimation. While the IIOT devices used are small, low-cost, and rugged to stand for harsh environment, they do have their limitations such as the raw CSI data are very noisy, which can lead to low performing models when directly fed to machine learning algorithms. To address this challenge, our previously developed statistics pattern analysis (SPA) framework [4][5] is used to extract robust and predictive features from raw CSI data. The SPA-based features are then used to develop robust machine learning models for continuous MC estimation in woodchips. In other words, SPA turns messy big data into so-called smart data, which can then be used for accurate sensing. The machine learning techniques studied in this work include artificial neural networks (ANN), support vector machines (SVM), and Gaussian process regression (GPR). We conducted systematic study of these nonlinear regression approaches, and demonstrated that, when combined with SPA, provide a potential solution for real-time woodchip MC estimation for the pulp and paper industry. Specifically, the experimental validation results show highly accurate prediction of more than 20 MC levels in a wide range. In addition to the excellent performance, the proposed approach is also economic and non-destructive, making it a potential technology for industrial implementation in the pulp and paper industry.

[1] R. Govett, A PRACTICAL GUIDE FOR THE DETERMINATION OF MOISTURE CONTENT OF WOODY BIOMASS A Practical Handbook of Basic Information, Definitions, Calculations, Practices and Procedures for Purchasers and Suppliers of Woody Biomass.

[2] D. Halperin, W. Hu, A. Sheth, and D. Wetherall, Tool Release: Gathering 802.11n Traces with Channel State Information, *ACM SIGCOMM Comput. Commun. Rev.*, 2011.

[3] W. Yang, X. Wang, S. Cao, H. Wang, and S. Mao, Multi-class wheat moisture detection with 5GHz Wi-Fi: A deep LSTM approach, in *Proceedings - International Conference on Computer Communications and Networks, ICCCN*, 2018.

[4] Q. P. He and J. Wang, Statistics pattern analysis: A new process monitoring framework and its application to semiconductor batch processes, *AIChE J.*, 2011.

[5] J. Wang and Q. P. He, Multivariate Statistical Process Monitoring Based on Statistics Pattern Analysis, *Ind. Eng. Chem. Res.*, 2010.

633025 Gradient-Weighted Class Activation Mapping (Grad-CAM) Based Explanations for Process Monitoring Results from Deep Neural Networks

Abstract:

In the chemical process industries, high product quality with maximum efficiency and process safety are the critical issues that require effective process monitoring. A fault in a process results in the deviation of process state variables from its normal operating condition. The effect of the fault propagates along the process and disturbs stability of the system. So, it is essential to detect them quickly on their occurrence. To tackle this problem, we motivate to develop the methodology which fulfils the aspect of high accuracy, novel explainability and to make effective use of the multivariate time series process data.

Recent developments in sensor techniques and storage technologies have provided a significant technology push for developing next-generation data driven process monitoring methods [1]. Hence deep learning methods have attracted much attention for fault detection. Deep learning methods like Deep Belief Network (DBN), Convolutional Neural Networks (CNN) and, Hierarchical Deep Neural Networks (HDNN) have been used for fault detection [2] but they lack in providing explanations for their decision. Hence, the eXplainable Artificial Intelligence (XAI) method Integrated Gradients was used by [3] to provide explanations for the results generated from deep learning models. But the Integrated Gradient suffer from one shortcoming that it analyses the samples one by one without accounting for the fact that the data is a time series. In the paper, we develop the XAI method which accounts for the time series nature of the process monitoring problem.

Recently, a new XAI method called Grad-CAM has been developed targeted at providing explanations for time series data in terms of activation maps for CNN [4]. Grad-CAM helps to explore the spatial information that is preserved inside convolution layers. The extracted information from CNN layers helps to understand the important feature of an input data to make classification decision. Grad-CAM uses k feature maps of height h and width u of the last convolution layer, as it contains high level semantics and detailed spatial information. For each feature map A^k, we obtain an importance weight w_k^c associated with specific class c. The w_k^c is obtained by computing the gradient of final class output y^c with respect to the feature map A^k i.e., $\partial y^c / \partial A^k$. It is further globally averaged over width u and height h. This step is followed by computing a weighted sum of feature maps A^k of class c using w_k^c. The final heatmap is obtained by applying the *ReLU* operation on the weighted sum to emphasize only positive values. This heatmap highlights the features which are important in making a final decision of the network for class c. In this paper, we use the Grad-CAM method for the explanation of process fault diagnosis.

In this paper, we exploit the real time explainable CNN methodology that classifies the normal and faulty state of the process. The proposed methodology consists of a 2D Convolution layer that is trained to solve a fault detection problem. Then, Grad-CAM is used on a trained model to understand which part of input data contributes more to detect a fault [5]. The contribution of different features visualized using the generated heatmap for each sample helps to identify the key factors in fault occurrence. The performance of the proposed methodology demonstrated on simulated time series classification datasets which contain different input variables and categorical outputs. The CNN was trained and heatmaps were generated using Grad-CAM that provide an accurate explanation about the predicted outcome. In this paper, we will describe the proposed methodology as well as results obtained from the simulated dataset which effectively makes use of the time series dimension to provide intuitive explanations for outputs. The results from fault diagnosis on the Tennessee Eastman challenge problem will also be demonstrated [6].

References

[1] P. Park, P. D. Marco, H. Shin, and J. Bang, "Fault Detection and Diagnosis Using Combined Autoencoder and Long Short-Term Memory Network," *Sensors*, vol. 19, no. 21, p. 4612, Oct. 2019, doi: 10.3390/s19214612.

[2] H. Wu and J. Zhao, "Deep convolutional neural network model based chemical process fault diagnosis," *Comput. Chem. Eng.*, vol. 115, pp. 185–197, Jul. 2018, doi: 10.1016/j.compchemeng.2018.04.009.

[3] V. Pakkiriswamy and R. Srinivasan, "An Explainable Artificial Intelligence Based Approach for Interpretation of Fault Detection Results from Deep Neural Networks," presented at the 2020 Virtual AIChE Annual Meeting, 2020, [Online]. Available: https://www.aiche.org/academy/conferences/aiche-annual-meeting/2020/proceeding/paper/375g-explainable-artificial-intelligence-based-approach-interpretation-fault-detection-results-deep.

[4] R. R. Selvaraju, M. Cogswell, A. Das, R. Vedantam, D. Parikh, and D. Batra, "Grad-CAM: Visual Explanations from Deep Networks via Gradient-based Localization," *ArXiv161002391 Cs*, Dec. 2019, doi: 10.1007/s11263-019-01228-7.

[5] R. Assaf and A. Schumann, "Explainable Deep Neural Networks for Multivariate Time Series Predictions," in *Proceedings of the Twenty-Eighth International Joint Conference on Artificial Intelligence*, Macao, China, Aug. 2019, pp. 6488–6490, doi: 10.24963/ijcai.2019/932.

[6] J. J. Downs and E. F. Vogel, "A plant-wide industrial process control problem," *Comput. Chem. Eng.*, vol. 17, no. 3, pp. 245–255, Mar. 1993, doi: 10.1016/0098-1354(93)80018-I.

630074 A Deep Learning Vision System for Classification of Manufacturing Defects

Abstract:

Manufacturing lines for the assembly of pharmaceutical autoinjectors are highly automated and rely on visual inspection stations to efficiently and safely process a large volume of syringes. As pre-filled syringes enter the inspection line, there are a variety of subtle defects in the flange (e.g., chips, cracks) as well as other processing flaws (e.g., missing components) that can make the assembly unfit for use or otherwise problematic for continued processing. Currently, there is a highly-engineered inspection algorithm that is manually tuned for identifying defects from camera images. The existing approach is able to robustly detect important defects, however, a relatively large percent of rejected products are in fact false positives (e.g., cosmetic defects, lighting irregularities). In this work, we have developed a convolutional neural network approach for classifying defects in syringes with comparable sensitivity and higher selectivity than the previous algorithm. We found that preprocessing and noise reduction were critical for training meaningful models from available images. Additionally, we highlight a series of infrastructure challenges that were overcome during image acquisition and algorithm deployment to transfer our work to and from a good manufacturing process (GMP) inspection line. We also illustrate the importance of algorithm-architecture co-design when developing machine learning models in order for the solution to be practically deployed on the available edge computing device (i.e., the vision processor). This work highlights the use of machine learning in conjunction with traditional computer vision to classify defects beyond what can be identified by expert intuition.

633005 Magnesium Oxychloride Formation Kinetics and Enhanced Water Stability for Sustainable Building Materials Applications

Abstract:

Magnesium oxychloride (MOC) is a ceramic material with significant fire resistant properties and growing potential as a light-weight, structurally strong alternative building material. Despite many advantages, MOC has limitations as a construction material, particularly water stability. While researchers have explored solutions to these shortcomings, fundamental understanding of this material is still urgently needed. Thus we have investigated the fundamentals of the formation cure reaction and mechanisms of water stability enhancement reactions. The formation cure reaction kinetics for magnesium oxychloride 5-phase was monitored from 35 to 55°C using time-resolved quantitative x-ray diffraction and differential scanning calorimetry (DSC). The reaction was characterized as a two-step process: dissolution of magnesium oxide into a gel state followed by crystallization of magnesium oxychloride. Assuming first-order kinetics for both MgO dissolution and MOC crystallization, a kinetic model predicts 42.4 kJ/mol and 26.1 kJ/mol for dissolution and crystallization activation energies respectively. Alternatively, the Avrami nucleation and growth model was fit to DSC measurements predicting diffusion controlled, one-dimensional growth with an activation energy of 72.4 kJ/mol, accounting for both dissolution and crystallization.

For the water stability enhancement, two methods of water stability enhancement have been explored: chlorartinite formation from CO_2 exposure and phosphoric acid addition. The conversion of magnesium oxychloride to chlorartinite by CO_2 forms a protective, semi-insoluble chlorartinite layer on the surface of the magnesium oxychloride crystals, which improves water stability. Phosphoric acid (2.5 to 10 wt. %) was added to the MOC slurry before the cure reaction. Additions of 2.5 wt. % and above had positive impacts on the water stability, preserving ~50 wt. % crystalline MOC after the water stability test, but has a significant impact on the MOC formation kinetics. This is achied via. the formation of an amorphous phase on the MOC crystal surface that contains structural motifs related to insoluble MgHPO4·3H2O (newberyte) and Mg2P2O7·3.5H2O (magnesium pyrophosphate) phases. Results from this work are significantly impactful for the use of MOC in more sustainable alternatives for conventional materials in residential and commercial building applications.

624619 Optimization of TEG Dehydration Process in Natural Gas Processing Under Metamodel Uncertainty

Abstract:

Optimization of TEG Dehydration Process in Natural Gas Processing under Metamodel Uncertainty

Rajib Mukherjee[1,2], Urmila M Diwekar[2*]

[1]Department of Chemical Engineering, The University of Texas Permian Basin, Odessa, TX 79762, USA

[2]Vishwamitra Research Institute, Crystal Lake, Illinois, 60012, USA

Abstract

Natural gas processing involves removal of acidic gases followed by dehydration that is mainly carried out through absorption in tri-ethylene glycol (TEG). The dehydration process is accompanied by the emission of volatile organic compounds, including BTEX. There are several process parameters associated with major equipment that may impact BTEX emission. In our previous work, a multi-objective optimization is performed for optimal operating conditions identification related to the process parameters to mitigate BTEX emission using data driven metamodeling using machine learning followed by optimization with metaheuristic technique [1]. The metamodel is generated with data obtained from process simulation carried out in ProMax® process simulator. The metamodels are created using limited samples which introduces metamodeling uncertainty [2]. Previous metamodel-based robust designs treated a metamodel as the real model and ignore the influence of metamodeling uncertainty. Thus, the performance of the resulting optimized process parameters may have shortcomings due to the lack of adequately accounting for the uncertainty introduced by the metamodel. In the present work, the bias of metamodel uncertainty has been addressed for parameter optimization. An algorithmic framework is developed to address the challenge of parameter optimization under metamodel uncertainties. We used the better optimization of nonlinear uncertain systems (BONUS) algorithm to solve the problem. BTEX mitigation is used as the objective of the optimization [3]. Our algorithm allows the determination of the optimal process parameters for BTEX emission mitigation from TEG dehydration process under metamodel uncertainty.

Keywords: Natural gas processing, TEG dehydration, BTEX mitigation, metamodeling uncertainty, support vector regression (SVR), BONUS algorithm

[1] Mukherjee, R., & Diwekar, U. M. (2021). Multi-objective Optimization of the TEG Dehydration Process for BTEX Emission Mitigation Using Machine-Learning and Metaheuristic Algorithms. ACS Sustainable Chemistry & Engineering, 9(3), 1213-1228.

[2] Zhang, S., Zhu, P., Chen, W., & Arendt, P. (2013). Concurrent treatment of parametric uncertainty and metamodeling uncertainty in robust design. *Structural and multidisciplinary optimization*, *47*(1), 63-76.

[3] Sahin KH, Diwekar U, M. (2004) Better Optimization of Nonlinear Uncertain Systems (BONUS): a new algorithm for stochastic programming using reweighting through Kernel density estimation. Ann Oper Res 132:47–68

633725 Mapping Environmental and Economic Analysis of Decentralized Cogeneration Energy Management Centers

Abstract:

In the midst of a global conversation on how to reduce GHG emissions, Ontario's electrical grid stands out due to its extensive use of low carbon-intense processes such as nuclear and hydropower that acts as baseload energy sources. However, for peak electricity demands that require a quick ramp up of production, natural gas (NG) turbines are deployed [1], which causes the carbon intensity of Ontario's electrical grid to vary hourly [2]. Another intensive and permanent Canadian energy requirement is heating. Currently, it is provided mostly by NG -fired boilers [3]. The pursuit for better NG utilization is the motivation for this study, which proposes a design of combined heat and power energy management center (CHP-EMC).

CHP-EMC is composed of an NG-powered internal combustion engine connected to an electrical generator (NG-CHP), a hot water tank storage, an NG boiler, and a cooling tower to keep the water temperature within acceptable bounds. EMC is connected to a water loop similar to the ones used in district heating (DH) by a heat exchanger and responds to a community's heating demand using an hourly lumped-demand approach. The simulations are developed in TRNSYS, which is a capable heat transfer and fluid dynamic simulator [4]. The storage tank uses a built-in dynamic model whilst the rest of the components operate in pseudo steady-state. These components are connected by pipe models that account for time delay and flowrate.

The equipment presence and its sizes are decided by minimizing the natural gas consumption while simultaneously maximizing thermal and electrical generation. The optimization strategy used is a black-box approach using modeFrontier software [5]. This optimization and data visualization software can be coupled with any other software that uses input/output files -such as TRNSYS- and it executes several simulations guided by the optimization strategy and its constraints. For this work, classic Multi-Objective Genetic Algorithm (MOGA) is chosen for design variables (including the presence of specific components) and the pareto points are analyzed in more detail. TRNSYS simulation has internal simple controllers that take care of schedule and operational constraints. The result of the optimization aims to create an EMC flexible enough to endure deviations from expected demand profiles.

The emissions of proposed EMC designs are compared to the status quo (electricity supplied by electrical grid and thermal demand supplied by NG boilers) for a year of operation. Total emissions are calculated for each possible value of gCO_{2eq}/kWh to better represent possible scenarios of carbon intensity. The results are summarized in an easy-to-read emissions vs technology map that defines the conditions under which the EMC is less pollutant than status quo (SQ). The optimal design and selection of the EMC (vs. SQ) system is strongly dependent on the carbon intensity of the electric grid during peak-load times, the electric demand profiles, and the cooling demand profiles over the course of its operation. These are best characterized by the combination of greenhouse gas emissions per kWh during peaking operation, the ranges of heating and electricity demand over the course of the year (the maximum minus the minimum), and the kurtosis of the demands.

Following this analysis, an economic evaluation of both systems is carried out, demonstrating the cost of carbon avoided (CCA). CCA is a metric that compares CO_2 and cost of two technological strategies and calculates the cost per ton of GHG emissions potentially reduced [6]. These results are compared to possible carbon tax values, in a similar way to carbon intensity. A second map, cost vs technology, is then developed for easy visualization of the impact of carbon taxes on EMC and SQ operation costs.

The goal of this work is to apply the benefits of decentralized EMCs to reduce GHG emissions from both heating and electricity generation. The results are generalizable and can be applied to any location due to its methodology in representing carbon intensity and carbon tax.

References

[1] IESO, "Power Data," 30 04 2021. [Online]. Available: https://www.ieso.ca/power-data. [Accessed 30 04 2021].

[2] L. Pereira and I. D. Posen, "Lifecycle greenhouse gas emissions from electricity in the province of Ontario at different temporal resolutions," *Journal of Cleaner Production,* vol. 270, no. 122514, 2020.

[3] S. Canada, "Primary heating systems and type of energy," 03 04 2021. [Online]. Available: https://www150.statcan.gc.ca/t1/tbl1/en/tv.action?pid=3810028601. [Accessed 28 04 2021].

[4] TRNSYS, 2019. [Online]. Available: http://www.trnsys.com/index.html. [Accessed 30 04 2021].

[5] ESTECO, "modeFrontier," 2021. [Online]. Available: https://engineering.esteco.com/modefrontier/. [Accessed 30 04 2021].

[6] K. Gillingham and J. H. Stock, "The Cost of Reducing Greenhouse Gas Emissions," *Journal of Economic Perspectives,* 2018.

627430 Development of an Interactive Software Tool for Designing Industrial Solvent Recovery Processes

Abstract:

Abstract

In 2017 the chemical industry was reported to be the world's second-largest manufacturing sector (United Nations Environment Programme 2019). Solvents are used in most operations in chemical and pharmaceutical industries as a reaction medium, selective dissolution and extraction medium, and dilution agents. Thus, there is a sizable amount of solvent waste generated due to process inefficiencies such as inefficient mixing, and inappropriate process equipment being utilized. There are several options for solvent waste disposal, however, the three most common are on-site disposal, off-site disposal, and incineration (J. Raymond, Stewart Slater, and J. Savelski 2010). Even though these methods of solvent waste disposal carry considerable negative environmental impacts they are still widely used since they are readily available and provide an easy method to dispose of large quantities of waste. Solvent recovery is typically not used because of the potential difficulties in achieving a required purity to meet the stringent guidelines imposed on the products in pharmaceuticals and specialty chemical industries as well as the additional infrastructure and investments needed. To this end, this problem needs to be studied carefully by involving all aspects in regard to capital needs, environmental benefits, comparison to traditional disposal methods, while achieving the required chemical purity.

Thus, designing a solvent recovery method is a multi-objective problem focused on minimizing capital investments of the new separation technologies while maximizing the amount of solvent recovery matching the desired purity requirements. This task is further complicated because there is a lack of a guidance system to assist professional engineers in the modeling and design of solvent recovery processes. Such a guidance system or tool that could predict the value of solvent recovery and reuse for industry users, can generate an environmentally friendly trend to shift away from using incineration, on-site disposal, and off-site disposal as primary solvent waste handling methods. As seen in Figure 1A, not only will this tool provide a greener alternative, but also aid in solvent recycling into the chemical plant, which is economically favorable. To this end, if we can develop a user-friendly tool that allows engineers to easily access solvent recovery options, we can provide an economical and environmentally favorable strategy.

To account for all possible waste streams a robust framework is needed. This consists of a maximal process flow diagram that encompassed multiple stages of separations and multiple technologies within those stages. Together this process flow diagram develops the superstructure that provides multiple technology pathway options for any solvent waste stream. We established the different stages in the superstructure as Solid Removal, Recovery, Purification, and Refinement by reviewing the literature on existing materials recovery and separation processes (Yenkie, Wu, and Maravelias 2017; Biegler 1997; Geankoplis 2003). Common technologies are placed in different stages depending on the component they can separate in terms of their physical and chemical properties. A comprehensive chemical database is created using sources, such as the Design Institute for Physical Properties (DIPPR), to store all relevant chemical and physical properties ("DIPPR" 2012). Using General Algebraic Modeling Systems (GAMS) to model the various separation units and stream flows as a multi-objective optimization problem, the amount of solvent recovered can be optimized while minimizing the environmental impacts and total cost of operation (Chea et al. 2020; Yenkie et al. 2016). With the GAMS code as the backbone, a Graphical User Interface (GUI) is created to provide a user-friendly tool to the chemical industry. Figure 1B shows how the GUI connects the superstructure modeled in GAMS model, the chemical database, and user inputs to allow for anyone in the industry to find an optimal technology pathway for their solvent waste.

The GUI will allow for a user to input chemical species and parameters that are specific to the process that is being observed. By connecting MATLAB to the GAMS code, it is possible to model a solvent recovery process solely in the GUI with these input parameters. This tool reports economic, chemical, and environmental factors for the user to consider. The economic factors include an annualized cost breakdown, utility costs, consumable costs, labor costs, and overhead costs. These costs can be seen for the whole selected recovery pathway with details of the individual technologies. The chemical factors report the amount of solvent the predicted pathway can recover and its purity level. The environmental factors show the estimated emissions from the technology pathway and compare these emissions against those for incinerating the waste stream. Thus, this tool is a powerful option for those in the industry who want to look for a cost-effective way of recycling their solvents from the waste streams.

Keywords: Solvent Recovery, Recycling, Graphical User Interface

Acknowledgments

This research was supported by the United States Environmental Protection Agency's Pollution Prevention (P2) Program (NP96259218). The authors thank the Department of Chemical Engineering at Rowan University for their continued assistance with acquiring the tools required for this research.

References:

Biegler, Lorenz. 1997. *Systematic Methods of Chemical Process Design.* 1st ed. Prentice Hall PTR.

Chea, John D., Austin L. Lehr, Jake P. Stengel, Mariano J. Savelski, C. Stewart Slater, and Kirti M. Yenkie. 2020. "Evaluation of Solvent Recovery Options for Economic Feasibility through a Superstructure-Based Optimization Framework." *Industrial & Engineering Chemistry Research* 59 (13): 5931–44. https://doi.org/10.1021/acs.iecr.9b06725.

"DIPPR." 2012. The American Institute of Chemical Engineers. March 29, 2012. aiche.org/dippr.

Geankoplis, Christie J. 2003. *Transport Processes and Unit Operations.* 4th ed. Englewood Cliffs, NJ: P T R Prentice-Hall, Inc.

2010. Raymond, Michael, C. Stewart Slater, and Mariano J. Savelski. 2010. "LCA Approach to the Analysis of Solvent Waste Issues in the Pharmaceutical Industry." *Green Chemistry* 12 (10): 1826–34. https://doi.org/10.1039/C003666H.

United Nations Environment Programme. 2019. "Global Chemicals Outlook II - From Legacies to Innovative Solutions - Synthesis Report (2019)." Global Chemicals Outlook II. United Nations Environment Programme. https://wedocs.unep.org/bitstream/handle/20.500.11822/27651/GCOII_synth.pdf?sequence=1&isAllowed=y.

US EPA, OEI. 2013. "TRI Data and Tools." Data and Tools. US EPA. March 3, 2013. https://www.epa.gov/toxics-release-inventory-tri-program/tri-data-and-tools.

Yenkie, Kirti M., WenZhao Wu, Ryan L. Clark, Brian F. Pfleger, Thatcher W. Root, and Christos T. Maravelias. 2016. "A Roadmap for the Synthesis of Separation Networks for the Recovery of Bio-Based Chemicals: Matching Biological and Process Feasibility." *Biotechnology Advances* 34 (8): 1362–83. https://doi.org/10.1016/j.biotechadv.2016.10.003.

Yenkie, Kirti M., Wenzhao Wu, and Christos T. Maravelias. 2017. "Synthesis and Analysis of Separation Networks for the Recovery of Intracellular Chemicals Generated from Microbial-Based Conversions." *Biotechnology for Biofuels* 10 (1). https://doi.org/10.1186/s13068-017-0804-2.

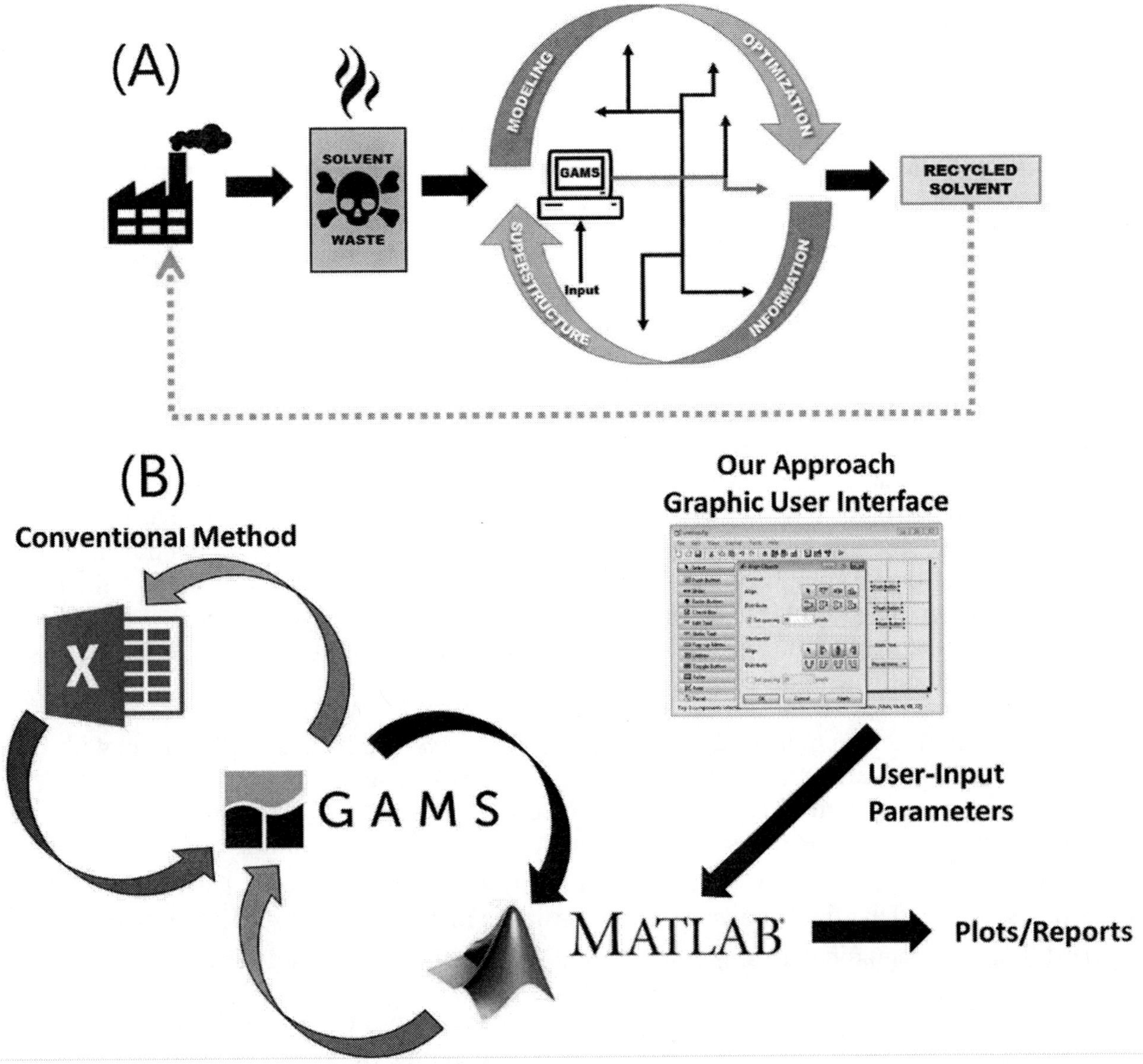

Figure 1(A) Proposed Process for Modeling Solvent Recovery for use in an Industrial setting.
(B) Flow chart of Information Processing of the Proposed Computational Software Tool

625002 High Flux CO2 Selective Membranes for Renewable Natural Gas and CO2 Capture

Abstract:

Reducing carbon emissions is key to preventing further climate change and in correcting the damage that has already been done. Compact Membrane Systems is developing a carbon dioxide selective membrane designed to operate at low pressure with minimal pretreatment. Results from two applications will be discussed: removal of carbon dioxide from flue gas and upgrading of biogas to renewable natural gas.

According to the World Biogas Association, anerobic digestors for biogas production can reduce greenhouse gas emissions by 10% in 2030. Separating carbon dioxide from methane is a key challenge for the renewable gas industry. Biogas is produced largely on farms or small waste centers, not at large centralized plants. While there are existing technologies to separate methane from carbon dioxide, they require multiple stages and high pressures, and are sensitive to process poisons. An efficient and cost-effective solution is needed to bring down barriers that previously prevented small scale operations from transforming their biogas into usable fuel. CMS has developed a membrane aimed at allowing users to capture, separate and upgrade existing biogas streams that would otherwise be flared or released into the atmosphere.

Results from the first field demonstration of Optiperm™ *biogas* will be shared. Working with Energy Projects Manager Matt Steiman at the Dickinson College Farm, a small spiral wound module system has been installed. The system takes raw biogas feed directly from the digester with no pretreatment. The pilot is intended to demonstrate successful operation under field conditions for an extended period of time. Optiperm™ *biogas* technology will be tested for a period of at least 2 months to demonstrate stability in the field and validate performance measured in the CMS laboratory. Optiperm™ *biogas* upgrading streams from 60% to 90% methane in a single stage.

Finally, initial results for utilizing the CMS membranes to capture carbon dioxide from flue gas, including the effect of operating conditions and initial system modeling, will be shared. This application is earlier in development, but results thus far have shown a very high flux membrane that could result in substantially lower capture costs than the current state of the art.

AICHE Fall 2021

Optimizing Energy Efficiency of Ammonia Production via Electrochemical Reaction and Haber-Bosch process.

Gbemisola Ojo and Kyle Camarda
[1]The University of Kansas, Department of Chemical & Petroleum Engineering
4132 Learned Hall, 1530 W 15th Street, Lawrence, KS 66045

Ammonia (NH_3) synthesis is crucial to the production of fertilizers, and accounts for nearly 2% of total global energy consumption. Therefore, finding methods to produce ammonia employing only renewable energy and sustainable feedstocks is paramount. This project uses dynamic simulation to design a 100% renewable ammonia synthesis plant, using only air and water as raw materials. Energy is supplied from a solar source, which is supplemented with industrial batteries. The intermittent nature of solar energy production requires the development of a dynamic production profile for the plant.

In this work, an electrolytic cell powered by a time-varying solar source is employed for hydrogen production. Air is separated via a membrane to provide a nitrogen source, and then the Haber-Bosch reaction is used to produce the product. The degrees of freedom in this problem are the number of batteries purchased, as well as the dynamic production and charging schedules. To determine the optimal production profile, a mixed integer non-linear programing model is formulated and solved. Results demonstrate how varying levels of production and energy storage usage correspond to different levels of energy availability, and show the feasibility of completely sustainable ammonia production, using no raw materials other than water and nitrogen, and creating zero emissions.

621410 Teaching Process Data Analytics and Machine Learning

Abstract:

Teaching chemical engineers to be effective in data analytics is as important today than in any time in the history of the discipline, and advances in sensor technologies, machine learning, and associated software have enabled applications that were not formerly possible. This abstract describes experiences with teaching process data analytics and machine learning over the last few years, including in (1) a new chemical engineering elective course in which the students come from chemical engineering, mechanical engineering, and engineering management, (2) a professional education course cotaught by chemical and mechanical engineering faculty whose learners primarily come with chemistry and chemical engineering backgrounds and many years of industrial experience, and (3) an undergraduate chemical engineering concentration in process data analytics with courses coming from a mixture of multiple disciplines including chemical engineering, computer science, economics, management. The overview will describe the challenges in teaching data science to chemical engineers and scientists and engineers working in the chemical and biotechnology industries, and strategies for overcoming those challenges.

622120 Teaching Big Data Science to Undergraduate Students in the University of Iowa

Abstract:

In this presentation, we will provide an overview of our course development and our experiences in teaching big data science and analytical skills to the undergraduate students in chemical engineering in the University of Iowa. These courses include: (a) Machine Learning (ML) and Artificial Intelligence (AI) course; (b) statistical analysis course; and (c) thermodynamic course. They are designed to be offered to students from freshman to junior years. We found that in general, the chemical engineering students are not getting used to use programming to analyze the data, in part because many courses they are taking focus more on the lab experiment while the analysis of the lab results is elementary and often done in Excel or by calculator. However, the students are willing to learn new analytical skills, and provided with well-designed teaching materials and problems, they can grasp the big data skills quickly. While the outcome of our experience has been encouraging, we also found there is clear lag in the field of chemical engineering to train our next generation with big data tools. This lag is part is due to the institutional tradition that emphasizes on the lab experiments and often deterministic way of thinking (with less consideration of probabilities), and in part due to the slowness in re-design the curriculum to reflect new development and ease access of ML and AI. We will provide examples to support the findings above, share the lessons we learned, and provide thoughts for next steps.

623878 Textile: Tutorials in Experimentalist Interactive Learning

Abstract:

Background:

Confocal microscopy is a common preclinical imaging modality to study a range of tissue features. Fluorescent image microscopy can provide insights into cell density, morphology, and cell-cell relationships. In our lab, we use immunofluorescent staining to investigate changes in cell and microenvironment features in response to development, injury, and treatment in the brain. We have developed image processing pipelines that threshold, segment, and quantify cell features. More recently, we have incorporated data science-enabled analysis and techniques into our imaging workflow. While the lab is comprised of researchers who are wet-lab experimentalists and data-driven analysts, only the second group has formal data science training. Wet-lab experimentalists can significantly benefit from having data science literacy to apply data science methodologies to their imaging datasets. However, there is no formal training regime to train wet-lab experimentalists in lab-specific data science methodologies. To address this training gap, we developed Tutorials in EXperimentalisT Interactive LEarning (TEXTILE) for data-science enabled fluorescent image processing.

TEXTILE is a module-based, semi-linear educational system developed with laboratory-based learning in mind. Modules combine data science fundamentals with specific biological imaging processing techniques. For example, we pair a data science module about data management with an image processing module on quantitative cell morphology characterization. The modular format allows us to train students on managing experimental data and pair that with quantifying features of cells that are relevant to their specific project of interest. Additionally, we designed TEXTILE such that the learning process is adaptable to starting skill level. Learning paths are based on entry knowledge but allow students to pick and choose their modules based on interest or previous experience. For example, someone with ample coding knowledge can skip the introductory Python modules and begin with the experimental image processing modules. Finally, we designed TEXTILE for a variety of degree levels. As a laboratory at a large public university, we engage with undergraduate and high school students, in addition to master's thesis and doctoral thesis students. We accomplished degree-level accessibility through the semi-linear system that assumes no prior knowledge in our research area. Since we already use TEXTILE to train high school students participating in laboratory research that have no previous knowledge of the subject topic, we aim to include more local high school students and expand TEXTILE for science education and outreach as well as in-laboratory education.

Methods:

TEXTILE is composed of topic-specific modules. Each module is designed around a data science topic, lasts an hour and a half, and contains four sections: the pre-module activity, main module, post-module reflection, and feedback form (Figure 1a). The instructor sends the pre-module activity to students for self-driven learning a week before teaching the main module. The pre-module's purpose is to prime students for the module's main lesson with a high-level activity. For example, in our data management module, students analyze a grocery shopping trip and decide where to store specific categories of food and perform a space evaluation on how much room in each storage location the food will take up.

The main module aims to teach the bulk material in digestible lectures broken up by interactive sessions. The main module's detailed structure consists of six parts: introduction presentation, interactive session one, main presentation workshop, interactive session two, concluding presentation, and a reflection session (Figure 1b). The introduction presentation introduces the module's topic, outlines the structure the main module will follow, and leads into the first interactive session. Interactive session one guides the student through an exploration of the module material. For coding-based modules, instructors lead interactive sessions with Jupyter Notebook. During interactive session one, the instructor walks students through the code's basics as students code along. After the students have written the code, the instructor begins the main presentation, which has two purposes: (1) to increase the students' connection between the pre-module activity and the module introduction and (2) to dive deeper into the technical aspects of the first interactive session. Once students understand the technical elements of the main module, they begin interactive session two, where they apply technical details of the lesson to further explore the code written in the earlier interactive session. After students have an opportunity to explore, the instructor gives the concluding presentation, which includes wrapping up the main takeaways, emphasizing the module's motivation, and thanking students for their participation.

Finally, the main module ends with a prompt for students to reflect on what they have learned and how their opinions about the module's topic have changed since the pre-module activity. The post-module reflection is completed individually by each student a day or two after the main module. The purpose is to provide a student-led review activity that encourages them to explore content from the main module on their own time to solidify concepts and build confidence. After the whole module is completed, students are sent a feedback form through Google Forms. With the feedback form, students reflect on their growth and the impact of each component of the module. The feedback is then implemented back into the module for the next cohort of students.

Results:

In the summer of 2020, the first cohort of students went through six TEXTILE modules of paired data science and image processing techniques used within the lab. The first cohort had 15 students: six high school students, four undergraduates, and five graduate students. The first cohort was taught live by a Ph.D. student from the lab's data-driven analysis team. The topics covered included GitHub and version control, introduction to data management, introduction to experimental data management specific to the lab, intro to image processing, thresholding, image segmentation, cellular morphology, and quantifying cell shape features. By the end of the first cohort, all students completed every TEXTILE module and the high school and undergraduate students developed individual projects using protocols learned over the summer. They each created Jupyter Notebooks for a self-led project with a brain cell morphology lit review, data management plan, and image processing pseudo-code. Many students from the summer 2020 cohort chose to continue researching with the lab and are currently developing these projects further.

In fall 2020, a small cohort of two undergraduate students explored the TEXTILE modules asynchronously. The students watched a recording of the main module and participated in asynchronous discussions with previous students and a Ph.D. student-instructor via Slack. We also provided these students with published image data that no one previously analyzed for cell morphology features. The students are currently completing retroactive analysis on this data using techniques learned from various TEXTILE modules.

For summer 2021, TEXTILE is expanded to include three modules on machine learning and two additional modules on experimental image acquisition and manual cell counting, incorporating student feedback from 2020. The main takeaways from student feedback included developing a formalized mentoring process for student projects and teaching via a hybrid live and asynchronous format.

Conclusions:

TEXTILE is a successful data science educational system for teaching data science techniques commonly used in experimental research labs. The TEXTILE system works for protocol training of fellow lab members, onboarding new students, and science education outreach. The combination of interactive sessions with instructor-led presentations gives students ample opportunity to work hands-on with real data from the lab in a guided environment. Students from the first cohort had the opportunity to develop their cell morphology research projects. In contrast, students from the second cohort took the lessons learned from TEXTILE and applied them for retroactive analysis of previously published data. The TEXTILE system is built to be robust and adaptable so that other laboratories can follow the methodology to develop training modules for their own data science techniques.

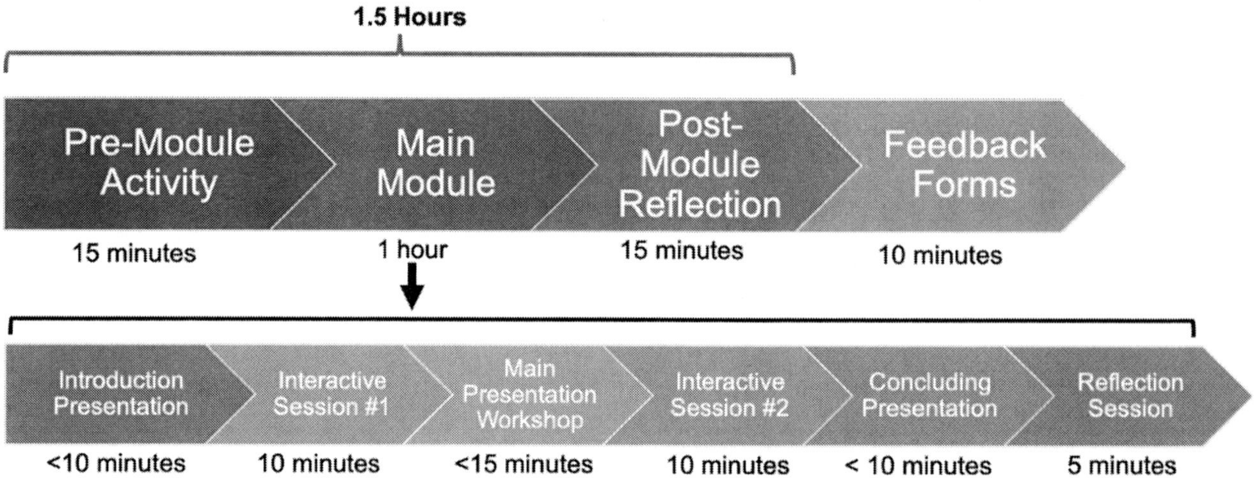

Figure 1. (a) Top row: TEXTILE process composed of four parts – pre-module activity, main modules, post-module reflection, and feedback forms. (b) Bottom Row: The six sections of the main module - introduction presentation, interactive session #1, main presentation workshop, interactive session #2, concluding presentation, reflection session.

622481 Teaching Artificial Intelligence to Chemical Engineers: Experiences from a 35-Year-Old Course

Abstract:

Given the recent success of artificial intelligence (AI) in a wide variety of applications, it has become quite clear that chemical engineering students need to become familiar with applied AI techniques [1]. There is great interest in the academic community in developing a course to address this important need. I have been teaching such a course to chemical engineering undergraduate and graduate students since 1986 [2], first at Columbia for three years, then at Purdue for about twenty years, and back again at Columbia now. It was mostly on *symbolic AI* in the early years, with an emphasis on *knowledge-based expert system* concepts and techniques, for various applications in chemical engineering such as design, diagnosis, control, optimization, and process safety. From the 1990s onwards, as machine learning techniques started showing promise, my course evolved to include topics such as clustering, neural networks, statistical classifiers, graph-based models, and genetic algorithms. The current version at Columbia has the first one-third of the course on *symbolic AI* and the rest on machine learning-based *numeric AI* concepts and techniques. It is mainly a project-based course where the students develop systems using AI techniques for chemical engineering applications. In the early years, the students programmed in AI languages such as LISP and OPS5, but now using the easier tools such as Python and TensorFlow. In this presentation, I will discuss my experiences in teaching this course, including the syllabus and the projects we have done over the decades.

Reference:

1. Venkatasubramanian, "The Promise of Artificial Intelligence in Chemical Engineering: Is it here, finally?", *Perspective article, AIChE Journal,* 65(2), pp. 466-478, 2019.

- Venkatasubramanian, V. "Artificial Intelligence in Process Engineering: Experiences from a Graduate Course," *Chemical Engineering Education*, Fall Graduate Education Issue, November 1986, pp. 188 – 192.

630903 Integrated Design and Model Predictive Control of Multiscale Systems Using a Multi-Fidelity Bayesian Optimization Approach

Abstract:

Design and control are interconnected activities that need to be performed simultaneously so that designs with a large degree of operational flexibility can be readily identified [1]. In practice, however, design is typically tackled first (with little-to-no considerations of the dynamics), which can lead to heavily constrained processes with few degrees of freedom available for control. While there is a vast body of literature available on the topic [2]–[4], a tractable and unified framework for optimal integrated design and control (IDC) has remained elusive. This type of framework is needed to enable the systematic design of next-generation (so-called "flexible") manufacturing and energy systems including smart grid technology [5], multiproduct or personalized chemical plants [6], and combined heat and power systems [7]. Although optimal IDC problems can be formulated as multistage stochastic programs (MSPs), they quickly become intractable when the following features are present [8]: (i) relevant dynamics and uncertainties occur on much shorter timescales than the lifetime of the system; (ii) uncertainties are present that are best described by continuous random variables with large variance; and/or (iii) important operational decisions are discrete (e.g., adaptive scheduling and unit commitment). The first two features relate to the multiscale nature of realistic IDC problems, which pose a major challenge for MSPs that suffer from the curse-of-dimensionality (i.e., exponential growth in number of decision variables as the control decisions can differ for each realization of the uncertainty) [9]. The third feature implies the system model is nonlinear with mixed-integer decisions such that a scenario approximation to the MSP is a very large-scale mixed-integer nonlinear program (MINLP) that is far beyond the capabilities of existing solvers, even when an equation-oriented system model is available.

Most previous work on optimal IDC focused on making tractable approximations to the MSP by relaxing features (i)-(iii) (see, e.g., [10]–[13] for several examples including substantially reducing the number of operational stages and uncertainty scenarios). Although these simplifications make the MSP more tractable, they degrade the accuracy of the IDC solution in exactly the ways that need to be captured in order to model the flexibility of the process. To address these challenges, in [14], we proposed a simulation-based optimization (SO) strategy that can tackle a general class of IDC problems consisting of an *outer* optimization over the relevant design variables and an *inner* stochastic simulation that evaluates the expected operating costs and constraints that appear in the outer problem. To reduce dimensionality of the outer problem, we rely on so-called decision rule (DR) approximations [8] to parametrize the recourse/control decisions in a way that scales much more favorably with respect to (i) and (ii) above. In particular, we presented a mixed-integer model predictive control (MIMPC) approach that is able to simultaneously handle scheduling and control decisions mentioned in (iii); this is a significantly more advanced DR than available alternatives that mostly consider PID controllers [15]–[17], with some exploring linear MPC [18], [19]. Due to the multiscale nature of the IDC problems of interest, we utilized Bayesian optimization (BO) that has become a popular tool for solving black-box optimization problems with objective and/or constraint functions that are prohibitively expensive to compute repeatedly and, therefore, must be evaluated as few times as possible.

In [14], we focused on conventional BO, which assumes only a single (computationally intensive) high-fidelity simulator is available. In IDC problems, however, we have access to several cheap approximations to this simulator such as the previously mentioned simplification methods in, e.g., [10]–[13]. As long as the lower fidelity approximations are reasonably correlated to the high-fidelity simulator, they can be used to cheaply eliminate "bad" designs and thus reserve the most expensive simulations for the most promising designs. To take advantage of these approximations here, we extend [14] by replacing traditional BO with a multi-fidelity BO (MFBO) algorithm [20]. To demonstrate its advantages, we apply MFBO to the design of a solar-powered building heating ventilation and air-conditioning (HVAC) system, with grid and battery support, under uncertain weather and demand conditions that vary at the hour scale for a year-long horizon. We show that MFBO repeatedly identifies more profitable designs than single fidelity BO under a fixed computational budget. In addition, we highlight that the choice and sequence of the lower-fidelity models has a strong impact on the rate of convergence of MFBO.

References

[1] V. Sakizlis, J. D. Perkins, and E. N. Pistikopoulos, "Recent advances in optimization-based simultaneous process and control design," *Comput. Chem. Eng.*, vol. 28, no. 10, pp. 2069–2086, Sep. 2004, doi: 10.1016/j.compchemeng.2004.03.018.

[2] L. A. Ricardez-Sandoval, H. M. Budman, and P. L. Douglas, "Integration of design and control for chemical processes: A review of the literature and some recent results," *Annual Reviews in Control*, vol. 33, no. 2. Pergamon, pp. 158–171, Dec. 01, 2009, doi: 10.1016/j.arcontrol.2009.06.001.

[3] Z. Yuan, B. Chen, G. Sin, and R. Gani, "State-of-the-art and progress in the optimization-based simultaneous design and control for chemical processes," *AIChE J.*, vol. 58, no. 6, pp. 1640–1659, Jun. 2012, doi: 10.1002/aic.13786.

[4] P. Vega, R. Lamanna de Rocco, S. Revollar, and M. Francisco, "Integrated design and control of chemical processes - Part I: Revision and classification," *Computers and Chemical Engineering*, vol. 71. Elsevier Ltd, pp. 602–617, Dec. 04, 2014, doi: 10.1016/j.compchemeng.2014.05.010.

[5] A. W. Dowling, R. Kumar, and V. M. Zavala, "A multi-scale optimization framework for electricity market participation," *Appl. Energy*, vol. 190, pp. 147–164, Mar. 2017, doi: 10.1016/j.apenergy.2016.12.081.

[6] M. C. Tang *et al.*, "Systematic approach for conceptual design of an integrated biorefinery with uncertainties," *Clean Technol. Environ. Policy*, vol. 15, no. 5, pp. 783–799, Oct. 2013, doi: 10.1007/s10098-013-0582-x.

[7] R. Madlener and C. Schmid, "Combined Heat and Power Generation in Liberalised Markets and a Carbon-Constrained World," *GAIA - Ecol. Perspect. Sci. Soc.*, vol. 12, no. 2, pp. 114–120, Apr. 2017, doi: 10.14512/gaia.12.2.8.

[8] A. Hakizimana, "Novel Optimization Approaches for Integrated Design and Operation of Smart Manufacturing and Energy Systems," Clemson University, 2019.

[9] M. V. F. Pereira and L. M. V. G. Pinto, "Multi-stage stochastic optimization applied to energy planning," *Math. Program.*, vol. 52, no. 1–3, pp. 359–375, May 1991, doi: 10.1007/BF01582895.

[10] I. E. Grossmann, B. A. Calfa, and P. Garcia-Herreros, "Evolution of concepts and models for quantifying resiliency and flexibility of chemical processes," *Comput. Chem. Eng.*, vol. 70, pp. 22–34,

Nov. 2014, doi: 10.1016/j.compchemeng.2013.12.013.

[11] P. Liu, Y. Fu, and A. Kargarian Marvasti, “Multi-stage stochastic optimal operation of energy-efficient building with combined heat and power system,” *Electr. Power Components Syst.*, vol. 42, no. 3–4, pp. 327–338, Mar. 2014, doi: 10.1080/15325008.2013.862324.

[12] D. K. Varvarezos, L. T. Biegler, and I. E. Grossmann, “Multiperiod design optimization with SQP decomposition,” *Comput. Chem. Eng.*, vol. 18, no. 7, pp. 579–595, Jul. 1994, doi: 10.1016/0098-1354(94)85002-X.

[13] T. Zhang, R. Baldick, and T. Deetjen, “Optimized generation capacity expansion using a further improved screening curve method,” *Electr. Power Syst. Res.*, vol. 124, pp. 47–54, Jul. 2015, doi: 10.1016/j.epsr.2015.02.017.

[14] N. A. Choksi and J. A. Paulson, “Simulation-based Integrated Design and Control with Embedded Mixed-Integer MPC using Constrained Bayesian Optimization,” in *Proceedings of the American Control Conference*, 2021.

[15] A. Flores-Tlacuahuac and L. T. Biegler, “Simultaneous mixed-integer dynamic optimization for integrated design and control,” *Comput. Chem. Eng.*, vol. 31, no. 5–6, pp. 588–600, May 2007, doi: 10.1016/j.compchemeng.2006.08.010.

[16] V. Bansal, J. D. Perkins, E. N. Pistikopoulos, R. Ross, and J. M. G. van Schijndel, “Simultaneous design and control optimisation under uncertainty,” *Comput. Chem. Eng.*, vol. 24, no. 2, pp. 261–266, 2000.

[17] M. Rafiei-Shishavan, S. Mehta, and L. A. Ricardez-Sandoval, “Simultaneous design and control under uncertainty: A back-off approach using power series expansions,” *Comput. Chem. Eng.*, vol. 99, pp. 66–81, 2017.

[18] P. Vega, R. Lamanna, S. Revollar, and M. Francisco, “Integrated design and control of chemical processes--Part II: An illustrative example,” *Comput. Chem. Eng.*, vol. 71, pp. 618–635, 2014.

[19] N. A. Diangelakis, B. Burnak, J. Katz, and E. N. Pistikopoulos, “Process design and control optimization: A simultaneous approach by multi-parametric programming,” *AIChE J.*, vol. 63, no. 11, pp. 4827–4846, 2017.

[20] K. Kandasamy, G. Dasarathy, J. Oliva, J. Schneider, and B. Póczos, “Gaussian Process Bandit Optimisation with Multi-fidelity Evaluations,” in *NIPS’16: Proceedings of the 30th International Conference on Neural Information Processing Systems*, 2016, pp. 1000–1008, doi: 10.5555/3157096.3157208.

627233 Adjustable Robust Optimization for the Planning Operations of Integrated Refinery-Petrochemical Site Under Demand Uncertainty

Abstract:

Facing the challenges of tight competition and environmental restriction from global market, petrochemical industry has been forced to seek all potential possible opportunities to improve the overall profit. Therefore, instead of considering planning optimization in refinery and petrochemical plant individually, a comprehensive standpoint of global enterprise optimization has received more attention than ever before [Li et al., 2016; Zhao et al., 2017; Uribe-Rodriguez et al., 2020]. The integrated refinery-petrochemical planning optimization provides a new route to achieve a higher profit margin by producing more valuable products, saving operating cost and raw material purchasing. Furthermore, ignoring the uncertainty in practical manufacturing, such as demand and sale price of final products, can reduce the total profit significantly. To improve the profitability of enterprise-wide refinery-petrochemical planning operations and avoid excess conservativeness derived from the worst-case based classical robust optimization, the framework of adjustable robust optimization under dynamic uncertainty set is proposed for a multi-period problem of integrated refinery-petrochemical site under uncertain demands.

Focusing on the operation planning of a large industrial refinery-petrochemical site which is consist of 24 and 15 processing units in each part, respectively, the nonconvex mixed-integer nonlinear programming (MINLP) is formulated. Complex process operation, such as the scheduling of ethylene cracking process and polymerization process are taken into account simultaneously. Binary variables are defined to denote the type of processed crude oil, purchase decision of raw crudes and selection of unit operation modes. Uncertainty lies in demand of final products and corresponding constraints are reformulated using robust counterpart, affinely adjustable robust counterpart, affinely adjustable robust counterpart with dynamic uncertainty set, respectively. The fluctuation and connection in demand of different periods are drawn from historic statistics. The problems are considered in three scenarios with time length varying in 5, 10, 15 periods respectively and solved by commercial solver BARON. Obtaining from the different models, the respective operational strategies are examined in diverse scenarios with random demand of final products. The computational results show that the method of affinely adjustable robust optimization coupled with dynamic set provides a best average performance on these sampling scenarios, following by the affinely adjustable robust optimization alone. The average profit of static robust optimization is the lowest but fluctuation between scenarios is smallest.

References

1. Li, J., Xiao, X., Boukouvala, F., Floudas, C. A., Zhao, B., Du, G., Su, X., & Liu, H., Data-driven mathematical modeling and global optimization framework for entire petrochemical planning operations, *AIChE Journal,* 62, 3020-3040 (2016)
2. Uribe-Rodriguez, A., Castro, P. M., Gonzalo, G.-G., & Chachuat, B., Global optimization of large-scale MIQCQPs via cluster decomposition: Application to short-term planning of an integrated refinery-petrochemical complex, *Computers & Chemical Engineering,* 140 (2020)
3. Zhao, H., Ierapetritou, M. G., Shah, N. K., & Rong, G., Integrated model of refining and petrochemical plant for enterprise-wide optimization, *Computers & Chemical Engineering,* 97, 194-207 (2017)

629298 A Novel Hybrid Algorithm for Scheduling of Multipurpose Batch Process

Abstract:

As the incremental market demands towards low-volume manufacture and various product categories, a growing number of researchers has been devoted to the investigation on batch processes [1,2]. The inherent flexibility of multipurpose batch process on processing route and shared facilities provides it wide application prospect in industry [2]. As one of the key managerial tools, process scheduling can greatly improve profit and save cost while saving makespan reflected on optimized production schedule. Mathematical programming is the most generally used method to address the batch process scheduling, being classified into sequence-based and time-grid-based models [3]. These formulations show good performance on handling scheduling with various processing features, such as variable processing time, storage policies and intermediate deliveries/demands. Continuous-time formulations [4,5] can effectively address multipurpose process scheduling in small-scale problems but often require large computational time or fail to converge to optimality for industrial-scale applications. The discrete-time formulation with tightening methods [6] is more computationally efficient and can handle intractable large-scale scheduling in reasonable time frames. However, in this approach, the optimal number of time points is not known a priori, which is often identified through the iterative procedure. As a result, the computational time is greatly increased.

Metaheuristics such as genetic algorithm (GA), tabu search and particle swarm optimization, have been proven capable to generate good scheduling solutions in a reasonable computational time even for large-scale batch processes. Many studies [7,8] confirmed that the GA shows good exploration of search space for multipurpose process scheduling attributed to its characteristics on inherent parallelism and strong global search ability. Also, GA can significantly reduce the computational time compared with mathematical programming where excessive CPU time is usually required for complicated scheduling problems. However, most existing GA proposed for multipurpose batch processes in literatures are less generalized and easy to prematurely converge to local optimality. Hybrid algorithm [9, 10] combining the advantages of multiple aforementioned methods can eliminate the limitations of single algorithm and show prominence on solving challenging large-scale problems.

In this work, we proposed a novel hybrid algorithm for scheduling of multipurpose batch process, which combines the strengths of GA on fast convergence and sequence-based mixed-integer linear programming (MILP) formulation on ensuring global optimum. The goal is to determine batching, allocation and sequencing of tasks eventually minimizing the Makespan under demand requirement. The hybrid algorithm comprises two stages: firstly, a new-designed GA is used to quickly generate feasible solutions, which do not require the iterative procedure to identify the optimal number of time points and hence significantly reduce the computational time. The obtained batching and sequence decisions are then used to fix some corresponding binary variables in MILP model. Then, the sequence-based formulation is quickly solved to optimality. Different fixing strategies of binary variables will be compared and identify the best one. In GA, a three-part encoding method is designed based on real value encoding way to direct batching and assigning determination and impact the processing sequence of tasks. Constraints concerning unit/storage capacity, material balance and product demand are handled in the decoding procedure to guarantee feasible solutions. We use elitist strategy and roulette wheel method [11] to select good individuals in each population. And a special crossover method was devised for the three-part real-value chromosome. We develop the sequence-based MILP formulation referring the model of Pan [12] and improve its efficiency by reformulating the big-M constraints concerning sequence and using less big-M terms to monitor inventory level of storage.

To evaluate the performance of the proposed method, four benchmark examples from the literature [5, 6] are solved using the proposed hybrid algorithm. Results show that the proposed algorithm can successfully generate global optimal solutions in all instances. It demonstrates general applicability and strong convergent performance. Compared with existing models, the hybrid algorithm can yield good quality solutions within shorter computational time. For example, in P9 and P11 of the Kallrath example, the method of Vooradi and Shaik [5] cannot converge to optimality in 40000s. By contrast, the proposed algorithm solves these two instances to optimality within 200 s, decreasing the CPU time over two orders of magnitude. In the most complicated instance P0, our approach reports a CPU time reduction by 37.5% compared to that of Velez et al. [6]. More importantly, the hybrid algorithm always generates better solutions than the designed GA.

References

1. Rakovitis N, Zhang N, Li J, Zhang LP. A new approach for scheduling of multipurpose batch processes with unlimited intermediate storage policy. *Chem. Sci. Eng.* 2019; 13(4):784-802.

2. Rakovitis N, Li J, Zhang N. An improved approach to scheduling multipurpose batch processes with conditional sequencing. *Aided Chem. Eng.* 2019; 46:1387-1392.

3. Harjunkoski I, Maravelias CT, Bongers P, et al. Scope for industrial applications of production scheduling models and solution methods. *Comput Chem Eng.* 2014; 62:161-193.

4. Vooradi R, Shaik MA, Gupta NM. Three-index model for Westenberger–Kallrath benchmark scheduling problem. *AIP Conf. Proc.* 2010; 1298:380–385.

5. Vooradi R, Shaik MA. Improved three-index unit-specific event-based model for short-term scheduling of batch plants, *Comput Chem Eng.* 2012; 43:148-172.

6. Velez S, Merchan AF, Maravelias CT. On the solution of large-scale mixed integer programming scheduling models. *Eng. Sci.* 2015; 136(2):139-157.

7. He Y, Hui CW. A binary coding genetic algorithm for multi-purpose process scheduling: A case study. Chem. Eng. Sci. 2010; 65(16):4816-4828.

8. Woolway M, Majozi T. A novel metaheuristic framework for the scheduling of multipurpose batch plants. Chem. Eng. Sci. 2018; 192:678-687.

9. Rakovitis N, Li D, Zhang N, et al. Novel Approaches for Energy-Efficient Flexible Job-Shop Scheduling Problems, *Eng. Trans.* 2020; 20: 823-828.

10. Lee H, Maravelias CT. Combining the advantages of discrete- and continuous-time scheduling models: Part 1. Framework and mathematical formulations, *Comput Chem Eng.* 2018; 116:176-190.

11. Beheshti Z, Shamsudding S. A review of population-based meta-heuristic algorithms, *Int J Adv Soft Comput Appl.* 2013; 5:1–35.

12. Pan M, Qian Y, Li XX. A novel precedence-based and heuristic approach for short-term scheduling of multipurpose batch plants. *Eng. Sci.* 2008; 63:4313-4332.

629370 Impact of Split Delivery in Minimizing LNG Procurement Cost

Abstract:

Historically, LNG has been traded via long term contracts between a producer and a customer. These long term contracts or LTCs, spanned over 15-25 years and provided both, the producer and customer with a stable supply of LNG. With the advent and growth of the LNG spot market in the last decade, many countries have been looking into methods to create a cost effective import portfolio considering both long term contracts and spot market. It is common practice in the LNG industry to load a single cargo for delivery to a single location supplied by a single seller. Since 2017[1], split delivery has also been introduced, which offers a new level of operational flexibility to LNG suppliers and importers. But with a split delivery, importers face a trade-off between increased operational costs and reduced cargo costs. Hence it is important for a buyer to create a procurement plan that takes this into account and optimizes overall procurement costs. In this study, we focus on the optimal procurement of LNG at the import terminal and study the benefits by including split delivery as an option.

In literature, LNG procurement for the customer has been studied primarily as a long term strategic decision with energy security as the priority while minimizing procurement cost. Shaikh et al. (2017)[2] modelled a natural gas import scheme for China, where they developed a mathematical model aimed at minimizing the import cost, transport distance, domestic and political instability associated with each supplier. Later, with a surge of spot market share in LNG trade, it was studied as the portfolio optimization problem that is determining the ratio of LTC and spot market in total imports. Kim and Kim (2018) [3] developed an optimal portfolio for Korea. In this study, they developed a two-step portfolio model mean-variance optimization model which determines the ratio of LTC and spot in imports followed by a linear programming model which chooses the optimal set of contract for LTC procurement. The benefits of split delivery in LNG supply chains have been discussed mainly from the producer's end. Mutlu et al. (2015) [4] considers the problem of developing an annual delivery program with least cost for a supplier. They included penalties for deviating from contractual terms, excess production and maintenance of ships and also a spot market revenue by selling excess LNG. Costs were compared with and without split delivery and they established the cost benefits of split delivery for a producer. Sheiktajian et al. (2020) [5] compared the costs of split and non-split delivery for an LNG manufacturer selling LNG via LTC and spot markets by considering deterministic and uncertain scenarios. Shahrukh et al. (2019) [6] were the first to develop an import strategy for an LNG importer under the terms of an ongoing long-term contract and a spot market available to minimize overall procurement costs for the short term (1-12 months) by proposing a mathematical formulation LNGPM. However they did not allow split delivery in their model. No model in existing literature can establish the benefits of split delivery at the customers end in the short term. In this paper, we seek to fill this gap by considering a scenario with split delivery and compare the costs to the scenario where no split delivery is allowed (LNGPM) to understand the various cost benefits due to splitting of delivery.

We have formulated the above problem as a mixed integer linear programming problem (MILP). In our formulation, we model the LNG available to the importer as a set of LTC and spot cargoes from which the optimal set having least overall cost has to be procured for satisfying the requirements of the importer. We consider multiple terminals at the importer's end, with each terminal having separate demand which must be collectively satisfied by the procured cargoes. We also consider limited berths/jetties and capacities of storage tanks at each terminal. For the cargos, we allow each cargo having its own volume, price, availability window and charter costs. In our model, we consider the option of splitting the delivery between the terminals available only to the LTC cargos and restrict the splitting of delivery between two terminals as seen in the real world. To prevent impractical solutions, the ratio of volume unloaded at each of the terminals is taken to be in reasonable limits . The objective function is to minimize the sum of cargo costs and transport costs under contractual stipulations and constraints based on the mentioned resources and obtain an optimal set of cargos to be procured.

To understand the benefits of split delivery, we present an illustrative example. The example involves an LNG importer having three regionally distributed terminals, having two LNG berths at each import terminal. The planning horizon is taken as 30 days. During the planning horizon, there are sixty spot cargoes available at different time windows. There are also 4 LTC cargos, each corresponding to a particular contract. Each contract has a minimum and maximum amount of LNG that can be procured from that supplier. Each LTC and spot cargo has a cargo volume, LNG price, travel times, and travel costs associated with each cargo. At each import terminal, capacity of storage tanks and a reorder level is specified. Travel time between all combinations of terminals is also defined in case of a split delivery. The illustrative example is solved for both scenarios and their results are compared. We see that there is an additional saving of 1.6 million USD by utilizing the split delivery option. We observe that using split delivery, we procure same number of LTC cargos, but slightly lesser volume of LNG. The other reason for savings is improved flexibility and better scheduling while ordering spot cargos. The additional travel cost incurred due to travel between terminals during a split delivery is offset by the savings in procurement cost. To reduce the effect of additional inventory at the end of the month, we solve the same scenario for 12 months on a rolling horizon basis to give a better comparison of the models. We see that by allowing split delivery, we are able to save 10.67 million USD (0.93%) when compared to the scenario without utilizing split delivery.

In the above example, we have considered benefits of only one customer who avails split delivery. There is a potential to reduce costs further by cooperating with other nearby customers and allowing split deliveries between import terminals of different customers to benefit both customers. In this regard, we have considered three levels of co-operation and compared the benefits among each level. We see that by co-operating and allowing split delivery there is always a reduction in cost seen for the customers. We will also report the observations from various scenarios discussed above.

References:

1. Dhul Qidah (2017, August 21). Qatargas, RasGas complete LNG co-loading for split delivery, LNG global. https://www.lngglobal.com/qatargas-rasgas-complete-lng-co-loading-for-split-delivery

2. Shaikh, F., Ji, Q., Fan, Y., Shaikh, P. H. & Uqaili, M. A. 'Modelling an optimal foreign natural gas import scheme for China'. J. Nat. Gas Sci. Eng. 40, 267–276 (2017).

3. Kim, J. & Kim, J. 'Optimal Portfolio for LNG Importation in Korea Using a Two-Step Portfolio Model and a Fuzzy Analytic Hierarchy Process'. Energies 11, 3049 (2018).

4. Fatih Mutlu, Mohamed K. Msakni, Hakan Yildiz, Erkut Sonmez, Shaligram Pokharel, 'A Comprehensive Annual Delivery Program for Upstream Liquefied Natural Gas Supply Chain'. European Journal of Operational Research (2015), doi: 10.1016/j.ejor.2015.10.031

5. Sheikhtajian S, Nazemi A and Feshari M. Marine Inventory-Routing Problem for Liquefied Natural Gas under Travel Time Uncertainty. International Journal of Supply and Operations Management, 7, no.1, pp 93-111 (2020)

6. Shahrukh, Mohd and Srinivasan, R. (2019). Evaluating the Benefits of LNG Procurement through Spot Market Purchase. Computer Aided Chemical Engineering, 46:1723–1728. doi: 10.1016/B978-0-12-818634-3.50288-5

634117 Keynote Talk: Toward Autonomy for Safe and on-Demand Biomanufacturing on Mars

Abstract:

Biomanufacturing holds promise as a versatile and economic means to transform in-situ resources into products that support long-term crewed space exploration and human mission to Mars. Bioproduction targets for Martian life-support systems primarily include pharmaceuticals, sustenance, and biopolymers that can be used in diverse additive manufacturing applications. A distinguishing feature of a space biomanufacturing factory compared to Earth-manufacturing factories is how versatile and robust a space factory must be to ensure crew survival in unknown environments and under mission uncertainties. In this talk, we will present our vision as to how the autonomy of a Martian biomanufacturing factory will require the safe, robust, and verifiable composition of deep learning, perception, mission planning, and optimal control in a feedback loop.

634118 Keynote Talk: The Roadmap to an Autonomous Chemistry Lab

Abstract:

Current approaches to chemical synthesis rely on time-consuming planning and labor-intensive manual synthesis, which is a rate-limiting step in the discovery of new functional molecules, a significant contributor to the overall cost of preclinical development, and clearly incompatible with the goals of autonomous discovery. In this talk, I will outline the components of a hypothetical autonomous synthesis laboratory designed to facilitate the development and manufacture of new small molecules. We will talk through the hierarchical process of computer-aided synthesis planning, starting from a list of molecules of interest and ending with an optimal schedule of automation-friendly action sequences for their sequential or parallel synthesis. Throughout, I will discuss our and others' progress towards this goal and where the primary technological barriers lay.

628373 Multi-Objective Supply Chain Optimization in Personalized Healthcare

Abstract:

Autologous Chimeric Antigen Receptor T cell therapies are a type of immunotherapy, where T cells are taken from a patient's blood and engineered to express the Chimeric Antigen Receptor (CAR). CAR T cells are then capable of targeting and killing cancer cells [1]. The two first commercial products have been approved in 2017 and 2018 by the U.S. Food and Drug Administration (FDA) and he European Medicines Agency (EMA) respectively for the treatment of B-cell lymphoma [2,3]. More recently, the U.S. FDA approved two more treatments for relapsed or refractory mantle cell lymphoma (2020) and for relapsed or refractory large B-cell lymphoma (2021) [4].

Unlike batch produced pharmaceuticals, CAR T cell therapies use patient cells as the starting material, placing the patient schedule in the center of the decision-making process. This unique 1-to-1 production and supply chain model translates into manufacturing lines and distribution nodes designed for and occupied by a single therapy, posing hurdles to the development of robust and responsive networks. Moreover, the predicted increase of patient numbers, the sensitive nature and short shelf-life of these therapeutics pose additional challenges to the coordination of manufacturing, storage and distribution [5].

In this work, we present a Mixed Integer Linear Programming (MILP) model for the optimal design of a supply chain network of CAR T cell therapies under different demand scenarios. The problem is formulated as multi-objective optimization aiming to minimize the average therapy cost (ATC) and the average return time of the therapies (ART). The preliminary results showcased in Figure 1 highlight the trade-off between the two objectives. The results are reported for two annual demand profiles of 200 (triangles) and 500 (squares) patients and two different manufacturing durations (TMFE). The base case scenario represents the current state of the art of 13 days of manufacturing (filled symbols), while a forward-looking scenario where the manufacturing time is reduced to 7 days is also considered (empty symbols).

We can observe that in the base case scenario an increase in the demand leads to an increase of the average therapy cost for the same ATR, which indicates that for prolonged manufacturing time the effect of economies of scale is not significant. On the other hand, in the forward-looking scenario it is observed that for long manufacturing times an increased demand may lead to facilities with capacity in order to satisfy all patients in time.

Future directions include: (i) investigation of scenarios with higher number of patients per year; and (ii) assessment of the effect of patient-specific manufacturing time on the supply chain network. The second point is particularly important, since the patient conditions at the time the cell sample is taken influence significantly cell doubling time and consequently the total manufacturing time required. The results will help identify bottlenecks in the overall manufacturing and distribution process, as well as the impact of a variable manufacturing time on the overall network. Finally, each scenario is simulated for different demand profiles in order to quantify the sensitivity of the patients scheduling for a given total demand and manufacturing time.

Acknowledgments

Funding from the UK Engineering & Physical Sciences Research Council (EPSRC) for the Future Targeted Healthcare Manufacturing Hub hosted at University College London with UK university partners is gratefully acknowledged (Grant Reference: EP/P006485/1). Financial and in-kind support from the consortium of industrial users and sector organizations is also acknowledged.

References

[1] B. L. Levine, J. Miskin, K. Wonnacott, and C. Keir, "Global Manufacturing of CAR T Cell Therapy," Mol. Ther. - Methods Clin. Dev., vol. 4, no. March, pp. 92–101, 2017.

[2] Novartis, "KYMRIAH Treatment Process, Dosing & Administration | HCP," 2018. [Online]. Available: https://www.hcp.novartis.com/products/kymriah/acute-lymphoblastic-leukemia-children/dosing-and-administration/. [Accessed: 24-Mar-2021].

[3] Kite Pharma, "First CAR T Therapy for Certain Types of Relapsed or Refractory B-Cell Lymphoma," 2018. [Online]. Available: https://www.yescartahcp.com/. [Accessed: 24-Mar-2021].

[4] A. Mullard, "FDA approves fourth CAR-T cell therapy". Reviews Drug Discovery 20, 166 (2021)

[5] R. Griffiths and M. Lakelin, "Successfully managing the unique demands of cell therapy supply chains." p. 9, 2017.

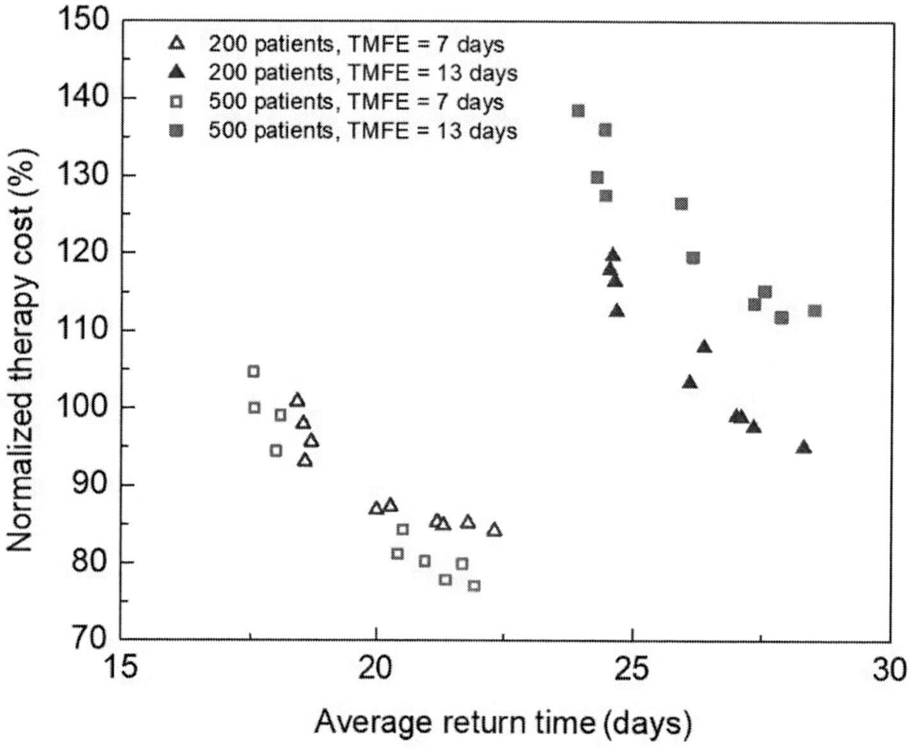

Figure 1: normalized therapy cost versus average return time of a therapy for two demand scenarios and two total manufacturing time excluding quality control.

630367 Robust Bioprocessing of Lignocellulose Using Microbial Tipping Points

Abstract:

Biofuels derived from lignocellulosic feedstocks have been shown to have near zero and in some cases net negative greenhouse gas emissions, driving public and private interest in producing lignocellulosic biofuels on an industrial scale. Anaerobic digestion (AD) is an established and scalable biological process for biogas production from a variety of feedstocks, including lignocellulosic biomass. AD uses a mixed microbial community adapted to the feedstock and operating conditions to produce biogas in four basic steps, hydrolysis, acidogenesis, acetogenesis, and methanogenesis. Biogas is on average 60 vol% methane and 40 vol% carbon dioxide and can be upgraded to renewable natural gas (RNG) by downstream carbon dioxide removal.

Lignocellulosic feedstocks are recalcitrant to hydrolysis, which is often the rate-limiting step for AD of this material. For other feedstocks, e.g. wastewater sludge, food waste, or manure, the rate-limiting step is considered to be methanogenesis. Because of this difference, the process conditions required for microbiomes to achieve high methane production rates during AD of lignocellulosic feedstock may have unique pH, temperature, residence time, and loading rate optima and tolerances. Hydrolysis-limited microbiomes may also have different sensitivities to perturbations of these conditions compared to more prevalent wastewater and manure systems. Understanding the sensitivities of AD to upsets or tipping points is necessary for operating a robust process, with robust defined as being able maintain the key process performance parameters without easily upsetting beyond recovery. A tipping point is defined as the process parameter value at which a biological community becomes unstable in terms of its key product(s) formation. This study proposes a method to define the tipping points for an AD microbiome fed senescent switchgrass. Within those tipping points the variation in process variables avoids irreversible reductions in system performance.

The environmental stresses evaluated in this study are temperature, pH, and overfeeding. These stresses were chosen as those likely to occur during full-scale operation, and that have been known to upset conventional waste digesters. Tipping points will be defined at similar thresholds as those used in inhibitor studies, with tipping points at 50% and 90% decline in methane production after the perturbation is applied. The use of two levels of failure is important as some stresses may not cause catastrophic failure within an operationally relevant range of stress magnitudes. The volatile fatty acid (VFA) concentration will also be monitored as a secondary variable used to identify and characterize a failing digester, as VFAs represent the key intermediates between the potential rate-limiting steps of hydrolysis or methanogenesis. For the pH and temperature tipping point experiments, both positive and negative changes will be evaluated, while for overfeeding only positive changes will be explored. The magnitude of each shock is provided in Table 1. Based on prior AD literature, lower feeding rates are unlikely to result in the dramatic and catastrophic failures that higher feeding rates and both higher and lower pH and temperature conditions can.

Each shock condition will be carried out in duplicate 1-L semi-continuous bioreactors and compared to duplicate controls. The digestate from a pilot scale anaerobic digester fed switchgrass at 55°C, 8.5 pH, and a 10-day retention time will be used to inoculate the 1-L bioreactors. After the shock is administered for a set time the bioreactors will be returned to the process conditions of the pilot reactor and monitored for three retention times for process recovery. Biogas production will be continuously monitored while the VFA concentration, feedstock conversion, and biogas composition are measured at 0.01, 0.1, 0.5, 1, 2, and 3 retention times.

This work aims to define the range of perturbation that a lignocellulosic AD system can experience without significant and irreversible loss of methane production. The method described allows for the sensitivity of a microbial system to be defined in terms of control variables. This then informs how precisely the system must be controlled to avoid biological tipping points leading to loss of product. Defining these tipping points and controlling within them will increase the operating efficiency of microbial systems and enhance the technical and economic feasibility of converting recalcitrant lignocellulosic feedstocks to biofuels.

Table 1. The process variable settings for each tipping point experiment

Process Variable	Settings
Overfeeding (g VS/L)	+10, +20, +30
Temperature (°C)	±10, ±20, ±30
pH	±0.5, ±1.0, ±1.5

624307 Keynote: Multi-Objective Optimization, State Estimation, and Advanced Control of a Semi-Batch Process for the Enzymatic Conversion of Lactose into Value-Added Products

Abstract:

Some of the main challenges experienced when scaling a laboratory enzymatic process to commercial scale include achieving optimal operations, performing online monitoring, and ensuring process controllability. Often times, laboratory-scale enzymatic processes correspond to simply proof of concept experiments that are carried out to characterize the reaction phenomena and develop preliminary process models. For example, in laboratory settings, ideal process conditions, such as negligible spatial variation in concentration, temperature or pH are much more achievable than at industrial scales[1]. As a result, when applying these models to industrial scales, a higher level of uncertainty and error is expected when compared to laboratory scale. Additionally, when developing these models, highly specific offline analytical methods are employed and relied upon due to their increased precision and accuracy over traditional online monitoring methods. However, in many industrial processes, soft sensors or estimation methods must be used online to infer process states from measurements of variables such as the oxygen uptake rate or carbon dioxide production rate. When using these methods, an additional level of uncertainty is introduced due to reliance on calibration ranges and model accuracy[2]. Thus, considering all these factors, it is often challenging to predict if laboratory scale processes can be feasibility scaled up. In this work, a process optimization, state estimation, and advanced control framework is introduced to determine the feasible scaled-up operations of a laboratory enzymatic process, considering a new enzymatic model for the conversion of waste lactose into a series of value-added products known as galactooligosaccharides (GOS).

The enzymatic model used in this work utilizes β-galactosidase to convert lactose into GOS products while distinguishing purely galactose derived di-, tri-, and tetrasaccharides from those containing a glucose group[3]. Due to the structural similarities of these isomers, online monitoring techniques are unable to isolate and uniquely distinguish the specific isomer concentrations. Instead, such techniques are only able to measure the combined concentrations of the isomers and provide one total concentration, thus posing a more challenging monitoring problem. Additionally, this specific model was chosen as it is currently at laboratory scale and could potentially be scaled to a commercial level to address the lactose waste problem plaguing small-scale cheese manufacturers[4].

Due to the many process uncertainties, such as raw feed cost and enzyme cost in this preliminary stage, conducting a robust and accurate economic analysis is challenging and would likely exhibit a wide distribution of economics, ultimately providing unreliable results. Instead, to examine the feasibility of scaling up this process, a multi-objective optimization is carried out considering 2 main factors: maximizing lactose conversion and GOS production rate as these will have the greatest effect on the process economics. Using the results of this multi-objective optimization, a trajectory-based model predictive controller (MPC) is implemented to guarantee that the process adheres to the optimum profile and is able to respond to simulated disturbances that are typical of industrial processes. Finally, a Moving Horizon Estimator (MHE)[5] is implemented to cope with the noise from the online measuring techniques and to decouple the combined isomer concentrations, providing insight into the specific compound concentrations in the reactor for optimal operations and control.

References

[1] Villadsen, J., Nielsen, J., Lidén, G. (2011). *Bioreaction Engineering Principles, 3.* Springer, Boston, MA.

[2] Stanke, M., Hitzmann, B. (2013). Automatic Control of Bioprocesses. In: Measurement, Monitoring, Modelling and Control of Bioprocesses, *Advances in Biochemical Engineering/Biotechnology, 1.* Springer, Berlin, Heidelberg.

[3] Schultz, G., Alexander, R., Lima, F. V., Giordano, R. C., Ribeiro, M. P. A. (2021). Kinetic Modeling of the Enzymatic Synthesis of Galacto-Oligosaccharides: Describing Galactobiose Formation. *Food and Bioproducts Processing, 127*, 1-13.

[4] Illanes, A. (2011). Whey Upgrading by Enzyme Biocatalysis. *Electronic Journal of Biotechnology. 14.*

[5] Alexander, R., Campani, G., Dinh, S., Lima, F.V. (2020). Challenges and Opportunities on Nonlinear State Estimation of Chemical and Biochemical Processes. *Processes, 8, 1462.*

631677 Keynote: Virtual-Engineering Software Framework for Integrated Biomass Conversion Modeling

Abstract:

This presentation covers the design and implementation of a software tool to systematically connect computational models of unit operations to simulate an integrated process of low-temperature conversion of biomass to fuel. This virtual engineering (VE) software was designed with the overarching goal of connecting unit models written in various programming languages and requiring different computational resources within a single, flexible framework. The models and features currently considered for the VE library include mechanistic models for pretreatment, enzymatic hydrolysis, and aerobic bioreaction; high-fidelity computational fluid dynamics (CFD) simulations for enzymatic hydrolysis and aerobic bioreaction; and the capability to perform techno-economic analyses (TEA) using Aspen Plus, a commercial software package. The CFD models require access to high-performance computing (HPC) resources, so in addition to handling multiple programming languages and interfaces, the VE software must also be capable of interacting with an HPC scheduler to submit, run, and post-process jobs.

Using the Python programming language, a new VE software package has been developed that contains functionality to manage the input-output communication between various unit models, schedule simulations to run on NREL's HPC and analyze those results, and interface with existing TEA software workflows. A Jupyter-notebook GUI was also created to solicit user input and provide documentation. In cases where multiple models for a particular unit-operation exist, selection between models is accomplished through a simple checkbox, with the appropriate inputs and outputs being parsed and converted seamlessly in the background. Each operation makes use of a different programming language, but the flow of information from pretreatment to enzymatic hydrolysis to bioreaction is managed with an intuitive, centralized file-communication strategy. In this talk, the programming approach and implementation details of the notebook are presented for multiple possibilities of the conversion process, including a demonstration of the ability to manage HPC resources. Additionally, an example of a sensitivity study of treatment parameters governing the overall conversion outcome is shown which highlights the ease of defining new problems using the VE Notebook workflow and leads into a discussion of ongoing work to enable outer-loop optimization studies.

630296 Comprehensive Planning of Annual Delivery Program for LNG Suppliers

Abstract:

The demand for natural gas has increased over the decade as it is the cleanest fossil fuel. However, it is economically infeasible to transport natural gas over long distances. Therefore, it is transported in liquefied form at cryogenic conditions using specialized ships known as LNG carriers. LNG supply chain planning is done on three levels namely strategic, tactical and operational planning. Strategic planning involves making long term decision that are effective over 20-30 years of duration. For e.g., selection of a contract portfolio, investment decisions associated with both liquefaction and regasification terminals, fleet composition and size etc. Tactical planning involves decision making which are effective over a horizon of 12 - 18 months. Creation of an Annual Delivery Program (ADP) by the supplier is an important tactical decision. Operational planning involves decision making which are effective over a horizon of 1 – 3 months. Rescheduling due to unforeseen events such as equipment breakdown or ship delays due to bad weather etc. are some examples of operational decisions. In this paper our focus is on tactical planning. LNG is procured by customers through Long Term Contracts (LTC) and spot market. LTC is an agreement signed between a supplier and a customer usually spanning up to 20-25 years of duration. So, with increasing demand, fleet sizes, customers around the world and other complicated decisions to make it becomes important for the supplier to plan deliveries to the customers at the accepted delivery dates throughout the supply period. To plan for deliveries for a coming year an Annual Delivery Program (ADP) is created by the supplier. ADP is the list of scheduled voyages which contains information about the long-term contract served, ships nominated to serve the contracts, regasification terminals nominated to receive delivery, date of ship loading etc. In this paper we present an LNG supply chain planning problem with an objective of creating an ADP to maximize the revenue generated by selling LNG in spot market and minimize the cost incurred by the supplier in delivering LNG to the long-term customers.

Currently, most of the literature present on creating ADPs is to satisfy the LTC demands only considering the inventory and berth management at the supplier's port. Rakke et al. (2011) developed a rolling horizon heuristic (RHH) for creating LNG annual delivery programs. A Mixed Integer Programming (MIP) formulation was proposed[1]. Stålhane et al. (2012) developed a construction and improvement heuristic (CIH) to create ADPs[2]. Their models allowed under supply of LNG and did not consider inventory and berth management at the customer's terminal. Mutlu et al. (2016) proposed a MIP model which incorporated split delivery for delivering LNG cargos. Commercial optimizers were not providing even a feasible solution to the MIP model in reasonable time. Small instances were tested but even those instances could not be solved to optimality in a 24-hour run time. So, they proposed a Vehicle Routing Heuristic (VRH) which gave cost effective solutions and outperformed commercial optimizers[3]. Sheikhtajian et. al. (2020) considered the model proposed by Mutlu et. al. (2015) and developed a model considering ship's speed as a fuzzy parameter. The aim of the study was to compare shipping expenses of split and non-split delivery strategies considering both deterministic and uncertain situations. They also proposed a meta heuristic as a solution method to solve the problem[4]. To summarize, some of the models in literature allowed under-supply neglecting inventory and berth management at the customer port, some models did not consider production and delivery of multiple grades of LNG, while others were not able to give optimal solution within reasonable time. We seek to address these gaps through our model.

In this paper, we present a Mixed Integer Linear Programming (MILP) model for delivering LNG shipments from the supplier to LTC customers and a regional spot market. The problem has one supplier and multiple customers spread around the world. The supplier produces and delivers multiple grades of LNG. Our objective is to create an ADP which maximizes profit for the supplier while considering ongoing long-term contracts and regional spot market. The supplier generates revenue by selling excess LNG in spot market. The cost incurred by the supplier in delivering LNG is the sum of transportation and penalty costs if it fails to abide by LTCs. The minimum and maximum production rates, storage capacities at the supplier and customer terminals, transportation time from supplier to customer ports and the slot wise demand at the customers terminal are known a priori. Our model allows over-supply to customers in a reasonable limit and penalizes the supplier for delivering more than the demand. In this paper, we also focus on the inventory and the berth management at the customer's port. The supplier is responsible for scheduling the heterogeneous fleet of ships. Our model explicitly accounts for loading of ships at the supplier's port, inventory of LNG at the supplier's and customer's terminal. The supplier sends excess LNG to the spot market over the planning horizon. Maintenance of ships is also incorporated in our model. The MILP model is demonstrated on various scenarios having different demands. We get an optimal ship schedule from our model which satisfies the inventory and berth constraints at the supplier and customer ports while fulfilling customers demand with reasonable over-supply. The MILP model for LNG delivery has been implemented in and solved using CPLEX. We believe that the results from this paper would help develop critical insights regarding the tactical planning aspects of the LNG supply chain from the perspective of both supplier and customer.

References

1. J. G. Rakke *et al.*, "A rolling horizon heuristic for creating a liquefied natural gas annual delivery program," *Transp. Res. Part C Emerg. Technol.*, vol. 19, no. 5, pp. 896–911, Aug. 2011, doi: 10.1016/j.trc.2010.09.006.

2. M. Stålhane, J. G. Rakke, C. R. Moe, H. Andersson, M. Christiansen, and K. Fagerholt, "A construction and improvement heuristic for a liquefied natural gas inventory routing problem," *Comput. Ind. Eng.*, vol. 62, no. 1, pp. 245–255, Feb. 2012, doi: 10.1016/j.cie.2011.09.011.

3. F. Mutlu, M. K. Msakni, H. Yildiz, E. Sönmez, and S. Pokharel, "A comprehensive annual delivery program for upstream liquefied natural gas supply chain," *Eur. J. Oper. Res.*, vol. 250, no. 1, pp. 120–130, Apr. 2016, doi: 10.1016/j.ejor.2015.10.031.

4. Sheikhtajian, Sanaz, Ali Nazemi, and Majid Feshari. "Marine Inventory-Routing Problem for Liquefied Natural Gas under Travel Time Uncertainty." *International Journal of Supply and Operations Management*7, no. 1 (2020): 93-111.

631079 A New Integrated Scheduling and Optimization Framework for Holistic Refinery Supply Chain Management

Abstract:

The holistic refinery supply chain management is critically important over the whole oil industry supply chain. It covers three highly correlated sub problems: the front-end sub problem, the refinery manufacturing sub problem and the multi oil-product pipeline distribution sub problem. According to Chima (2007), there is a need to ensure that each sector along the supply-chain can respond quickly to the exact demand of its downstream customers, protecting itself from problems with suppliers and buffering its operations from the demand and supply uncertainty it faces. Since the oil companies are aiming at maximizing the profit margin by weighing benefits versus costs throughout the entire refinery supply-chain. Thus, by coordinating the management and operations of the three sub problems along the holistic refinery supply-chain, it can greatly maximize the potential benefit margin. Although there exists a lot of works explicitly and deeply exploited and studied these sectors separately, only a few works addressed these sub systems along the refinery supply-chain simultaneously. Guyonnet et al. (2009) integrated the crude oil unloading, production planning, and distribution sub models on a tactical decision level by solving each part in a sequential push or pull manner, where the planning horizon is discretized into time periods, typically a day or a week. Siwi et al. (2018) proposed a strategical planning of the petroleum and petrochemical supply chain. However, systematic studies for the integrated scheduling and optimization of the holistic refinery supply chain are still lacking.

To minimize the total operating cost along the holistic petroleum refinery supply-chain, this work developed a new integrated scheduling and optimization framework covering the crude oil unloading, transferring, manufacturing, and pipeline distribution subsystems. Note that the specifications requirement for crude oil blending and oil product blending (where nonlinearity may occur due to the blending operations) must be satisfied. Meanwhile, operating rules such as tank can either receive or discharge oil product at a time has been considered, which further closed the gap between the modeling and the realistic problem (Yu et al., 2020). Also, the inventory/capacity limits, delivery feasibility constraints and oil product demands must be satisfied, which may result in a multi-stage and large-scale MINLP model. The efficacy of the developed integrated scheduling and optimization framework is demonstrated by an industrial-scale case study.

Keywords: Integrated scheduling and optimization, MINLP, Holistic refinery supply chain, Crude unloading and transferring, Refinery manufacturing, Pipeline distribution

631348 Integrated Load Shifting and Curtailment for Demand Response of Central Chilled Water Plants

Abstract:

The participation of industrial electricity users in Demand Response (DR) programs has increased in the past decade (FERC, 2019), motivated by the necessity of ensuring the electricity grid's generation-load balance and transmission reliability. Programs that promote load shifting and curtailment and the dispatching of distributed generation assets have become common products for independent system operators. While load shifting as a response to price signals has been widely studied in the past, the provision of load curtailment services from the viewpoint of the user started to gain attention in the last decade. Previous studies in this area considered processes such as cement milling (Vujanic *et al.*, 2012), aluminum smelters (Zhang and Hug, 2015) and air separation plants (Zhang *et al.*, 2015). In these works, there is a considerable variation in the load curtailment program (e.g., how far in advance the bids are placed and how far into the future they are implemented) and industrial process addressed, the combination of which results in different problem setups and solution approaches.

In this work, we study how load curtailment bids can be optimized and implemented along with load shifting and analyze its closed-loop performance under forecast uncertainty, specifically for large-scale central chilled water plants with Thermal Energy Storage (TES) used for District Cooling. First, an optimization framework is proposed that integrates the two problems of real-time load shifting (considering hourly day-ahead prices) and day-ahead load curtailment bidding and implementation. We perform a cost-benefit analysis that demonstrates the potential benefits of providing load curtailment under varying incentive profiles. Lastly, we analyze the effect of forecast uncertainty on the closed-loop performance and identify opportunities for integrating uncertainty into the optimization problem. The case-studies are performed with real data from the central chilled water plant located on the campus of the University of California, Davis.

References

FERC, 2019. 2019 Assessment of Demand Response and Advanced Metering, Technical Report.

R. Vujanic, S. Mariéthos, P. Goulart, M. Morari, 2012. Robust integer optimization and scheduling problems for large electricity consumers. Proceedings of the 2012 American Control Conference, 3108–13.

Q. Zhang, I. E. Grossmann, C. F. Heuberger, A. Sundaramoorthy, J. M. Pinto, 2015. Air separation with cryogenic energy storage: optimal scheduling considering electric energy and reserve markets. AIChE Journal 61 (5), 1547–58.

X. Zhang, G. Hug, 2015. Bidding Strategy in Energy and Spinning Reserve Markets for Aluminum Smelters Demand Response. In: Innovative Smart Grid Technologies Conference (ISGT), 2015 IEEE PowerEnergy Society, 1–5.

624503 A Web-Based Application for Chemical Production Scheduling

Abstract:

While a large range of models and methods have been proposed in the literature for chemical production scheduling, very limited tools are available for researchers to try and test these methods. In this work, we develop a web-based application to facilitate the generation and solution of chemical production scheduling problems. The application offers a graphical interface through which the user can define an instance using the state-task network representation (Kondili *et al.*, 1993). The connections between the states and tasks can be created using connectors and processing data (e.g., task processing times, conversion coefficients) can be entered through the interface.

Given additional user-defined parameters (e.g., horizon, optimality gap) and choice of the objective function (e.g., profit maximization, cost minimization, makespan minimization), the application then automatically generates the corresponding mixed-integer linear programming discrete-time model. In terms of solution methods, the user can choose among three options: (1) tightening constraints, based on a constraint propagation algorithm (Velez *et al.*, 2013); (2) reformulation based on the introduction of integer variables representing the total number of executed batches of each task (Velez and Maravelias, 2013); and (3) a discrete-continue algorithm which identifies good solutions fast based on a discrete-time model and then refines the solution based on a continuous time representation (Lee and Maravelias, 2018).

A data string representing the created network is generated automatically and submitted to the NEOS server (hosted at the University of Wisconsin-Madison). The user must provide their email address to retrieve the results. When the user submits the network through the web application, the NEOS server queues the relevant network data and based on the solution method(s) chosen, appropriate constraints are activated in the optimization model solved using the CPLEX solver in the GAMS platform. The user can access the model and solution statistics as well as a visual representation of the solution in the form of a Gantt chart along with the inventory profiles, which are zipped into a folder and emailed to the user.

References

1. Kondili, E., Pantelides, C.C., Sargent, R.W.H. (1993) A general algorithm for short-term scheduling of batch operations—I. MILP formulation. *Computers & Chemical Engineering*, **17**, 211–227.
2. Velez, S., Sundaramoorthy, A., Maravelias, C.T. (2013) Valid inequalities based on demand propagation for chemical production scheduling MIP models. *AIChE Journal*, **59**, 872–887.
3. Velez, S., Maravelias, C.T. (2013) Reformulations and branching methods for mixed-integer programming chemical production scheduling models. *Industrial Engineering and Chemistry Research*, **52**, 3832–3841.
4. Lee, H., Maravelias, C.T. (2018) Combining the advantages of discrete and continuous-time scheduling models: Part 1. Framework and mathematical formulations. *Computers & Chemical Engineering*, **116**, 176-190.

634119 Keynote Talk - Positive Power with Negative Emissions: Flexible Ngcc Enabled By Modular Direct Air Capture (DAC)

Abstract:

Natural gas power generation is a vital part of the U.S. energy system based on an abundant domestic energy source. In this talk a way to modify natural gas combined cycle (NGCC) plants such that they produce power with a negative carbon footprint will be described. To achieve this, we have integrated the natural gas combined cycle (NGCC) system with both conventional amine capture and direct air capture (DAC). This combination provides dispatchable operation of the combined system to deliver carbon negative electricity even when variable renewable energy (VRE) resources are scarce. During times of high VRE, the unit would reduce its electrical output and allocate more energy to the DAC system, removing more CO2 from the atmosphere driving the carbon footprint further negative. This requires manipulation of the DAC system operation to optimize steam use to enable maximum value to be derived from time varying pricing of electricity.

The NGCC plant will include conventional amine scrubbing to remove 97% of the CO_2 contained in the flue gas. The amount of DAC will depend on the balance of using steam from the heat recovery steam generation system to make power through the turbine system and its use for regenerating the DAC sorbent. The NGCC plant will also be able to ramp up and down maintaining a minimum load to allow for fast response to pricing signals. Overall this will allow for carbon negative electricity to be produced between 60-100% of the NGCC operating capacity with an additional penalty of 5-20% of the NGCC generating capacity, above that required for conventional capture, depending on how much steam is directed to the DAC plant.

634120 Keynote Talk - Machine Learning of Molecular and Materials Properties at the Low-Data Limit

Abstract:

The development of next-generation chemical manufacturing technologies will immensely benefit from robust, reliable, and accurate multiscale models that capture the underlying physical phenomena at multiple scales. A challenge, to this end, is to develop fast and accurate models of molecular and material properties using a combination of experimental and ab initio techniques. In this context, recent years have witnessed a burgeoning growth in the application of data science and machine learning for predicting such properties from large, homogeneous, and often computed molecular/material property datasets. Many properties, however, are hard to measure or compute; consequently, the available data in such cases is scarce, heterogeneous (that is available from across a variety of sources), and of differing fidelity. This talk will focus on data-driven model building for such "imperfect" datasets.

Using examples from molecule design and catalysis, this talk will focus on different methods, arising from process systems engineering, that can be applied to address the aforementioned challenges. First, **active learning** techniques that balance exploration of the molecule/material space and exploitation of the current model will be discussed to learn linear (generalized group additive) and nonlinear (graph convolutional neural network) property models. The second example will focus on the concept of **transfer learning** whereby information (model structure and features) from a model of a molecular property for which data is plentiful is subsequently leveraged ("transferred") while training a model of a related property for which data is scarce. Third, we will show that, by infusing a relatively large amount of low fidelity data via **multitask learning**, the thermodynamic properties of adsorption on catalytic surfaces can be modeled to a high level of accuracy using only small amounts of high accuracy data.

625584 Model-Based Investigation of Upstream CHO Cell Culture Process for Production of Monoclonal Antibodies with Desired N-Linked Glycosylation

Abstract:

N-linked glycosylation is post-translational protein modification that takes place during synthesis of monoclonal antibodies (mAbs) [1]. N-linked glycan composition or glycan index, including total afucosylation, high mannosylation, fucosylation, and/or galactosylation indexes, are often critical quality attributes (CQAs) of mAb based therapeutic proteins [2]. Researchers have found controlling the cell culture media components such as uridine, galactose, manganese, and ammonia concentration can tweak the glycosylation composition with the desired CQA levels [3-5]. During bioreactor operations, many process parameters or operating conditions can also affect protein glycosylation such as temperature, pH and dissolved oxygen [6-8]. However, the effect of process parameters on mAb glycosylation can vary significantly between different cell lines. To better understand the protein glycosylation process and further optimize the bioreactor operations, process modeling can be used to support quality by design (QbD) by correlating critical process parameters with critical quality attributes and further defining a design space for process operations.

We have investigated the effect of temperature and pH on protein glycosylation for a model Herceptin biosimilar (or Trastuzumab) production using a CHO cell culture cultivation in an upstream fed-batch bioreactor. Mechanistic kinetic model was developed based on the experimental data obtained under different temperature and pH combinations. The mechanistic model contains two parts, including an unstructured cell culture model and a structured intracellular model. For our unstructured cell culture model, viable cell density, glucose, ammonia, and protein concentrations are captured by empirical equations. In the structured model, the intracellular kinetics of biochemical reactions cascade of the N-linked glycosylation process are simulated. The kinetic model is used to understand the effect of temperature and pH on the bioprocess. Finally, feasibility analysis with surrogate-based adaptive sampling is applied to determine a bioreactor operation design space. In this combined experimental and computational study, we have determined how adjusting the bioreactor operating conditions can achieve critical quality attributes of the targeted product. Ongoing work will implement integration with a control system to achieve real-time process control for mAb glycosylation.

References

1. Akmal, M. A.; Rasool, N.; Khan, Y. D. J. P. o., Prediction of N-linked glycosylation sites using position relative features and statistical moments. **2017,** *12* (8), e0181966.
2. Szekrenyes, A.; Szigeti, M.; Dvorakova, V.; Jarvas, G.; Guttman, A., Quantitative comparison of the N-glycosylation of therapeutic glycoproteins using the Glycosimilarity Index. A tutorial. *TrAC Trends in Analytical Chemistry* **2020,** *122*, 115728.
3. Gramer, M. J.; Eckblad, J. J.; Donahue, R.; Brown, J.; Shultz, C.; Vickerman, K.; Priem, P.; van den Bremer, E. T.; Gerritsen, J.; van Berkel, P. H., Modulation of antibody galactosylation through feeding of uridine, manganese chloride, and galactose. *Biotechnol Bioeng* **2011,** *108* (7), 1591-602.
4. Kotidis, P.; Jedrzejewski, P.; Sou, S. N.; Sellick, C.; Polizzi, K.; del Val, I. J.; Kontoravdi, C., Model-based optimization of antibody galactosylation in CHO cell culture. **2019,** *116* (7), 1612-1626.

5. Ehret, J.; Zimmermann, M.; Eichhorn, T.; Zimmer, A., Impact of cell culture media additives on IgG glycosylation produced in Chinese hamster ovary cells. *Biotechnol Bioeng* **2019,** *116* (4), 816-830.
6. Villiger, T. K.; Scibona, E.; Stettler, M.; Broly, H.; Morbidelli, M.; Soos, M., Controlling the time evolution of mAb N-linked glycosylation - Part II: Model-based predictions. **2016,** *32* (5), 1135-1148.
7. Sou, S. N.; Sellick, C.; Lee, K.; Mason, A.; Kyriakopoulos, S.; Polizzi, K. M.; Kontoravdi, C., How does mild hypothermia affect monoclonal antibody glycosylation? *Biotechnol Bioeng* **2015,** *112* (6), 1165-76.
8. del Val, I. J.; Kontoravdi, C.; Nagy, J. M., Towards the implementation of quality by design to the production of therapeutic monoclonal antibodies with desired glycosylation patterns. *Biotechnol Prog* **2010,** *26* (6), 1505-27.

624781 Next-Generation Vaccines and Therapeutics: Towards Resilient Pharmaceutical Supply Chains

Abstract:

The promising clinical outcomes of genetic engineering have paved the way to curing life-threatening diseases, including cancers. Several innovative gene therapy drugs have already been launched over the last few years and this rapid growth is forecasted to continue with the U.S. Food and Drug Administration (FDA) anticipating the approval of 10-20 cell and gene therapy products per year by 2025 [1]. Approximately 70% of the cell and gene therapies in the clinical trials utilise viral vectors for gene transfer [2]. The versatility of viral vectors has also emerged in preventive healthcare, with their recent approval as vaccine platforms (e.g. Vaxzevria (AstraZeneca) and the Janssen COVID-19 vaccine) [3][4].

Given the high demand and limited manufacturing capacity, viral vector manufacturers are asked to make decisions on how to best utilise and expand the available manufacturing resources in the presence of numerous demand uncertainties. Some of the challenges are related to production and capacity planning under uncertain clinical trial outcomes and dosage requirements. Furthermore, manufacturers of viral vector-based COVID-19 vaccines tackle the hurdles of rapid and in-risk manufacturing, transitioning to large-scale deployment of novel vaccine products in record times. In this case, SARS-Cov-2 has clearly demonstrated the need for sophisticated tools that can assist responsive decision-making and support planning under uncertainty.

This work presents a novel mixed-integer linear programming (MILP) model for the design and optimisation of viral vector supply chains (Figure 1). An array of applications for viral vectors and their respective demand scales are considered, thereby capturing the emerging paradigm shift from *one-size-fits-all* manufacturing and distribution to targeted healthcare [5]. The framework relies on techno-economic analyses of well-established viral vector manufacturing protocols, which are modelled and simulated in SuperPro Designer. The formulated optimisation model identifies good candidate supply chain structures, capacities and production plans for multi-product multi-suite facilities, distribution plans and transportation flows. For this, we consider (i) a set of products and demand scenarios, (ii) capital and operating manufacturing expenditures for a range of production scales, as well as storage and transport costs for sensitive formulated products, (iii) candidate locations for network nodes, namely primary, secondary manufacturing and distribution centres. The resilience of the strategic decisions is then assessed under demand uncertainty through comparisons of network structures with respect to (i) cost, (ii), scalability and (iii) flow time.

References

[1] **U.S. Food and Drug Administration,** "Statement from FDA Commissioner Scott Gottlieb, M.D. and Marks, M.D., Ph.D., Director of the Centre for Biologics Evaluation and Research on new policies to advance development of safe and effective cell and gene therapies", https://www.fda.gov/news-events/press-announcements/statement-fda-commissioner-scott-gottlieb-md-and-peter-marks-md-phd-director-center-biologics, (2019).

[2] **Ginn, S., Amaya, A. K., Alexander, I. E., Edelstein, M., Abedi, M.R.,** "Gene therapy clinical trials worldwide to 2017: An update", *The Journal of Gene Medicine,* **20**(5), e3015, (2018).

[3] **European Medicines Agency,** "Vaxzevria (previously COVID-19 Vaccine AstraZeneca)", https://www.ema.europa.eu/en/medicines/human/EPAR/vaxzevria-previously-covid-19-vaccine-astrazeneca#overview-section, (2021).

[4] **U.S. Food and Drug Administration**, "Janssen COVID-19 Vaccine", https://www.fda.gov/emergency-preparedness-and-response/coronavirus-disease-2019-covid-19/janssen-

covid-19-vaccine, (2021).

[5] **Sarkis, M., Bernardi, A., Shah, N., Papathanasiou, M.,** "Emerging Challenges and Opportunities in Pharmaceutical Manufacturing and Distribution", *Processes*, **9**(3), 457, (2021).

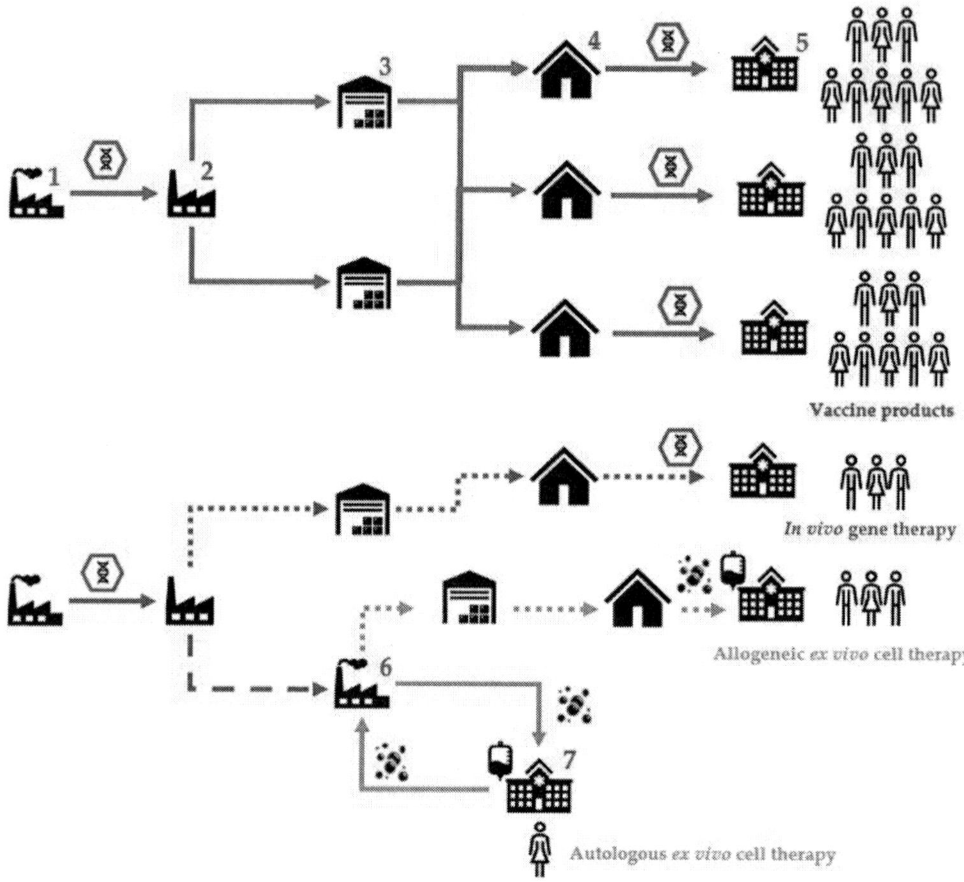

Figure 1: General schematic of supply chains for viral vector applications, comprising (1) primary manufacturing, (2) secondary manufacturing/fill-to-finish, (3) regional storage, (4) district storage, (5) hospitals/administration points as well as (7) specialised leukapheresis sites, (6) cell therapy manufacturing facilities.

624019 Systematic Decomposition & Evaluation of a Process Design Space for Monoclonal Antibody (mAb) Manufacturing

Abstract:

Three of the major trials within drug development and commercialization are phase 1 to 3 clinical trials. As these trials proceed product-process understanding increases and these are utilized via a feedback loop to the different development areas. The complex feedback loop consists of knowledge from input and output material properties (product related) and unit operations performance (process related). Monoclonal antibodies (mAbs) are of importance because they are proteins that are purposely engineered to treat various severe diseases for example cancer or hemophilia (Wolf Pérez et al., 2019). A typical pharmaceutical process for a specific active pharmaceutical ingredient (API) is batch-based because batch processes provide the flexibility of quality control at different intermediate steps and batch tracing inside the process boundaries and outside, that is, the product supply chain. Of interest, whether a continuous or batch process, is the evaluation of technologies that have a positive impact on process performance. Therefore, it is of interest to systematically evaluate relevant technologies for improving process performance and sustainability beyond high level "what-if analyses" to provide the foundation for advanced process understanding, design and optimization (Yang et al., 2020).

The mAb platform process to be investigated is pre-selected to be a batch process and therefore, any pre-selected flowsheet model for representation of the process is dynamic. Therefore, in analyzing upscale data for the platform process (lab and/or pilot), a systematic approach where multiple ideas associated with evaluating new technologies with respect to mass throughput and time execution is of importance. For the platform process, based on the large number of alternative placements of new technologies (integer decisions) and estimation of the input/outputs (continuous decisions) a systematic approach consisting of two layers of decisions is proposed. The decomposition layer addresses the integer decision problem, that is, placement of the new technology and the evaluation layer; having fixed the placement; addresses the continuous decision, that is, the expected gain in process performance, capacity and sustainability. This is performed using a simulation-based approach of material flow analysis with embedded non-linear process behavior.

The objective of this presentation is to present the systematic method for mAb process design space generation and screening. First, the problem and method will be presented. Second, the concept of discrete-event simulation will be introduced and how this can be used within the method for decomposition and evaluation of the design space. Finally, the method will be exemplified through an example of a promising technology for mAb manufacturing.

References

Wolf Pérez A. M., Sormanni P., Andersen J. S., Sakhnini L. I., Rodriguez-Leon I., Bjelke J. R., Gajhede A. J., De Maria L., Otzen D. E., Vendruscolo M. and Lorenzen N. 2014. In vitro and in silico assessment of the developability of a designed monoclonal antibody library. mAbs, 11(2), 388–400.

Yang, O., Qadan, M. & Ierapetritou, M. Economic Analysis of Batch and Continuous Biopharmaceutical Antibody Production: a Review. 2019. Journal of pharmaceutical innovation, 15, 182–200.

621647 Keynote Talk: Integrated Quality By Design in (Bio)Pharmaceutical Manufacturing

Abstract:

This presentation describes a Quality-by-Design approach for an accelerated process development workflow for the design of integrated (bio)pharmaceutical manufacturing systems from drug substance to drug product. This approach draws R&D and management into closer alignment and execution while defining product critical quality attributes, the operational design space, the manufacturing process, and the control strategy by integrated use of the ICH quality quartet – namely Q8(R2), Q9, Q10, and Q11. The workflow employs process intensification, micro-scale technologies, modular unit operations with integrated process control and monitoring systems, systems integration, and full plant automation. The presentation describes how to best develop and transfer knowledge between steps in the workflow via mechanistic models, data analytics, and machine learning. The approach is illustrated in more than a half dozen small- and large-molecule applications, including where process equipment was first designed digitally using first-principles models, then constructed and implemented experimentally, with the experimental results confirming model predictions. These applications include a wide variety of upstream, downstream, and formulation unit operations for small-molecule and biological drug manufacturing, as well as the design and operation of fully automated end-to-end manufacturing systems.

624003 Keynote Talk: An *in silico* approach for Monoclonal Antibody (mAb) Process Research & Early Development

Abstract:

Monoclonal antibodies (mAbs) are a type of biopharmaceutical that can be used to treat various diseases, for example, hemophilia and other serious diseases (Wolf Pérez et al., 2019). Improvements for the development of mAbs can be categorized as follows, product related improvements, for example, expression, solubility etc. and process related improvements, for example, novel flowsheets, new technologies (unit operations) etc. The synergistic effect of product-process decisions is also of importance because it addresses the integrated problem, that is, how can process design decisions be accounted for in the product design phase. Consider the example, given a specified mAb, purify the mAb using a chromatography column and selected eluent at a predetermined concentration. If the resulting aggregated pool volume for a finite time horizon for a pilot and/or industrial process requires a non-standard tank size that increases capital investment, how can this be a priori mitigated? If product and process design is performed sequentially then the design decision is unchanged however, if product-process design is performed integrated then, additional technologies can be evaluated for up-concentration. Therefore, the process is sized to the required scale and not oversized which has both capital investment and time-to-market implications.

Production planning-scheduling (PPS) is a field within operations research that solves the dynamic problem as to what to produce when (Harjunkoski et al., 2014). It can be performed using optimization-based methods and simulation-based methods, for example, discrete event simulation. The former differs from the latter because it provides an optimal solution for a set of decisions however, the latter provides the opportunity to deep dive into the details of the plan-schedule. Therefore, the symbiosis between the two approaches can be explained as follows. The former generates the optimal plan-schedule and the latter simulates how this is executed. In other words, the latter acts as a digital version of the process, a so-called digital twin. Therefore, we can use discrete event simulation to also evaluate the impact of different technologies and various design decisions on pharmaceutical processes, evaluate economic and sustainability impacts and improve decision making a priori to final investment (Yang et la., 2020).

The objective of this presentation is to present an *in silico*, systematic framework for multiscale process design and idea evaluation for biopharmaceutical products. A case study applied to mAbs will be presented. The method will be presented, and its application exemplified through idea evaluation related to new technologies for mAb production. Finally, the screening of the evaluation space using a combination of economic and sustainability metrics will be presented.

References

Harjunkoski I, Maravelias CT, Bongers P, Castro PM, Engell S, Grossmann IE, Hooker J, Méndez C, Sand G and Wassick J. 2014. Scope for industrial applications of production scheduling models and solution methods. Computers and Chemical Engineering, 62, 161–193.

Wolf Pérez A. M., Sormanni P., Andersen J. S., Sakhnini L. I., Rodriguez-Leon I., Bjelke J. R., Gajhede A. J., De Maria L., Otzen D. E., Vendruscolo M. and Lorenzen N. 2014. In vitro and in silico assessment of the developability of a designed monoclonal antibody library. mAbs, 11(2), 388–400.

Yang, O., Qadan, M. & Ierapetritou, M. 2019. Economic Analysis of Batch and Continuous Biopharmaceutical Antibody Production: a Review. Journal of pharmaceutical innovation, 15, 182–200.

622636 Developing a Workflow for Continuous Centrifugal Extraction

Abstract:

Liquid-liquid extraction is ubiquitous throughout API process development as a method to purify reaction mixtures. On paper, extraction is a straightforward, effective unit operation in which two immiscible liquids are mixed to extract specific components from one liquid phase to the other. In practice, extractions can pose problems as they are scaled up from development work to manufacturing scales. Traditionally, extraction in the pharmaceutical industry is performed in batch mode in stirred tanks, which can lead to long settling times and inefficient use of solvent. Furthermore, the presence of surfactants or the over-agitation of the two liquids may lead to emulsification, which significantly complicates the gravity-driven separation of the two liquids. Prolonged settling times increase the cycle time of the process, reducing overall throughput and may risk degrading reactive intermediates as the liquids remain in contact while settling.

Continuous extraction offers attractive, practical solutions to conventional batch extraction. First, continuous flow devices have smaller footprints compared to batch devices. Additionally, there are many types of flow contacting devices, such as mixer-settlers, packed or agitated columns, and membranes, among others, allowing flexibility for the contacting device to match the properties and needs of the two liquids in the process. One such continuous flow contacting device is a centrifugal extractor. These devices provide excellent mixing for the two liquids as they flow into the device while efficiently separating the two liquids via centrifugal force as they flow out of the device. Centrifugal extractors offer mixing and separation in the same unit to minimize the footprint, can separate liquids with small density differences, can tolerate liquids that tend to emulsify, and provide short contacting times for fast extractions or unstable intermediates. These advantages of centrifugal extractors make them attractive as options for continuous flow in the pharmaceutical industry.

This presentation describes the development work of a model system for continuous centrifugal extraction. We will describe the key features of the process, such as partition coefficient, number of equilibrium stages, and other relevant thermodynamic data. Additionally, we will highlight some of the key issues and parameters to address with scale-up and design of centrifugal extraction processes. Finally, we will detail AbbVie's general workflow to designing and developing continuous extractions with the goal of implementing robust continuous extractions at scale.

Eric Moschetta and Benjamin Rizkin are employees of AbbVie and may own AbbVie stock. AbbVie sponsored and funded the study; contributed to the design; participated in the collection, analysis, and interpretation of data, and in writing, reviewing, and approval of the final publication.

629625 A Scalable Continuous Reaction and Isolation Process for the Production of Sulfonyl Chloride Pharmaceutical Intermediates

Abstract:

Chlorosulfonic acid is essential for many pharmaceutical processes due to its ability to selectively chlorosulfonate aryl rings via an S_E2 mechanism (*1*). The chlorosulfonates are precursors to a wide range of biologically active molecules including sulfonylureas (*2*) and sulfonamides (*3*). The hazardous and corrosive nature of chlorosulfonic acid is challenging to address particularly during scale up for an industrial manufacturing process (*1*). The approach taken to address these challenges was to perform this reaction in two consecutive continuous stirred tank reactors followed by a continuous isolation. The implementation of continuous manufacturing can be advantageous to improve the safety, speed, product quality, and environmental footprint of challenging reactions (*4*). The safety advantages of transitioning this reaction to continuous manufacturing include a reduced amount of heated chlorosulfonic acid, the gradual release of toxic gases, and limited operator exposure. The transition to a continuous isolation step can greatly improve space-time yield as the slow addition of the final product mixture to the isolation vessel was determined to be an essential parameter for maintaining the critical quality attributes.

Following the development of preliminary reaction and quenching conditions, multiple 100-g scale flow experiments were performed using two continuously stirred round bottom flasks, each with a one hour mean residence time. Peristaltic pumps were used with a combination of C-Flex and PTFE tubing to transfer the reaction mixture between each vessel. The initial smaller scale flow runs revealed a need for process monitoring of CSTR levels so that flow rates could be adjusted as needed. Mettler Toledo scales were connected to a Python terminal via a serial port connection and a custom script was implemented so that changes in relative levels of the starting material and each round bottom could be detected and recorded in real time. The isolation step was performed with periodic rapid pumping from the quenching vessel into the filtration step to prevent excessive solid product accumulation.

A 500-g continuous manufacturing run was next performed with a 20-L jacketed reactor used for the precipitation step. A high-purity sulfonyl chloride product was obtained in moderate yield as determined by HPLC assay, ^1H NMR, and LCMS. Process monitoring data from the Python script revealed that the increased flow rate of the larger scale manufacturing run led to problems with C-Flex tubing compatibility and to subsequent variable flow rates and skewed residence times. Peristaltic pump heads for PTFE tubing were therefore purchased and implemented for all future flow experiments. Further batch-scale chemistry development elucidated that a higher number of equivalents of chlorosulfonic acid was needed to react with the sulfonic acid intermediate and that the initial work-up should be done at sub-ambient conditions. In order to improve green chemistry metrics, experiments were conducted minimizing the amount of water used during the isolation step. Lastly, the recommended process improvements were implemented in 100-g scale and 500-g scale continuous manufacturing runs.

Literature Cited

1. **Cremlyn, R.J.**, "Chlorosulfonic Acid: A Versatile Reagent," Royal Society of Chemistry,

Cambridge, UK (2002).
2. **Sen S., Ruchika, Kumar D., Easwari T.S., and S. Gohri,** "Therapeutic Aspects of Sulfonylureas: A Brief Review," *Journal of Chemical and Pharmaceutical Research,* **8** (12), pp. 121-130 (Dec. 2016).
3. **Kolaczek, A., Fusiarz, I., Lawecka, J., and D. Branowska,** "Biological Activity and Synthesis of Sulfonamide Derivatives: A Brief Review," *CHEMIK,* **68** (7), pp. 620-628 (2014).
4. **Baumann, M., Moody, T.S., Smyth, M., and S. Wharry,** "A Perspective on Continuous Flow Chemistry in the Pharmaceutical Industry," *Process Res. Dev.,* **24** (10), pp. 1802–1813 (2020).

626756 Target Polymorphic Form Development Via Continuous Combined Cooling and Antisolvent Crystallization Using Oscillatory Baffled Crystallizer

Abstract:

In recent years, continuous manufacturing has received much interest across the pharmaceutical industry, academia and regulatory bodies due to its potential to lower the cost of manufacturing while maintaining consistent quality[1-3]. The regulatory bodies, such as the U.S. Food and Drug Administration (FDA), have been encouraging pharmaceutical companies to adopt continuous manufacturing to ensure uninterrupted supply of high-quality drugs to patients[4]. One of the important links in continuous manufacturing is crystallization unit operation. Most often crystallization governs the critical quality attributes of active pharmaceutical ingredient (API) such as purity, crystal size distribution (CSD), yield, polymorphic form, morphology, etc. In continuous crystallization, the controlled state of operation helps in achieving consistent quality drug substance. Continuous cooling only crystallization on model compounds has been extensively studied[5-6], but a very few papers have been published on continuous combined cooling and antisolvent crystallization (CCAC). It is common in pharmaceutical industry to use antisolvent to improve yield, but the solvent composition can impact the polymorphic outcome of the drug substance[7]. Therefore, designing a continuous CCAC process to solve the challenges of API crystallization in a highly regulated environment is a complex engineering problem. In this work, continuous CCAC of a commercial drug substance, Atorvastatin calcium, via oscillatory baffled crystallizer will be presented.

Atorvastatin calcium (ASC) is a statin medication that is used as a lipid lowering agent and to prevent cardiovascular diseases. It works by inhibiting HMG-CoA reductase, an enzyme found in liver tissue that plays a key role in production of cholesterol in the body. With cardiovascular diseases being the number one cause of death in the world[8], the demand for the statin drugs is quite high, making it a high-volume drug substance and hence a suitable candidate for continuous manufacturing.

ASC is currently manufactured via batch CCAC, which is time intensive and gives wide CSD product at the end of the batch. To overcome this and other disadvantages of batch crystallization, continuous CCAC of ASC was studied in an Alconbury Weston Ltd DN-15 oscillatory baffled crystallizer (OBC) with the aim to improve productivity and CSD of the desired polymorph. An OBC has the advantage of high heat transfer rates and improved mixing that significantly brings down the crystallization time. It also has the advantage of spatial temperature distribution and multiple antisolvent addition points to control supersaturation and hence the crystallization process.

The desired polymorph of ASC form I is crystallized from the mixture of isopropyl alcohol (IPA) and water. First the solubility of ASC form I was found at different temperatures and solvent compositions. The solubility increased as the temperature increased but the solvent composition had a synergistic effect on the solubility. As the desired form is a trihydrate, the saturated API solution was prepared in a high IPA:water ratio and the final solvent composition had low IPA:water ratio.

Three different modes of batch CCAC process were studied to understand the effects of temperature and solvent compositions on the polymorphic outcome. The first mode was direct antisolvent addition in which the antisolvent added to the seeded saturated API solution. In the second mode, reverse antisolvent addition was carried out in which saturated API solution was added to seeded antisolvent.

In the third mode, distributed antisolvent addition was tested in which the antisolvent was split into two parts (AS-1 and AS-2). The saturated API solution was added in the seeded AS-1 and then the remaining AS-2 was added to system. The first mode did not yield the desired polymorph, but the second and the third modes gave the desired form. In the first mode, the IPA rich solution did not favor the desired form nucleation, whereas in the other two modes, the IPA concentration was much lower which favored the nucleation of the desired form.

In case of continuous crystallization, there is no direct or reverse antisolvent crystallization as the saturated API and antisolvent streams mix in the OBC tube at a point irrespective of which is added first. However, the ratio of IPA:water at the point of mixing of the two streams was found to have an impact on the polymorphic outcome and the operation in the OBC. The presence of seeds in either saturated API solution or antisolvent also impacted the polymorphic outcome. The splitting of antisolvent into two addition ports in the OBC was found to give the desired form. Through enhanced micromixing and increased heat transfer characteristics of the OBC, the continuous process was found to be 70-fold more productive compared to the batch process. The crystals obtained from continuous OBC operation had narrower CSD compared to that from a batch CCAC process.

Acknowledgement

Dr. Reddy's Laboratories Ltd. (DRL) is acknowledged for financial support provided through Purdue – DRL fellowship program.

Reference:

1. McKenzie, P., Kiang, S., Tom, J., Rubin, A. E. & Futran, M. Can Pharmaceutical Process Development Become High Tech? AIChE J. 52, 3990–3994 (2006).
2. Plumb, K. Continuous processing in the pharmaceutical industry: Changing the mind set. Chem. Eng. Res. Des. 83, 730–738 (2005).
3. Lee, S. L., O'Connor, T. F., Yang, X., Cruz, C. N., Chatterjee, S., Madurawe, R. D., Woodcock, J. Modernizing Pharmaceutical Manufacturing: from Batch to Continuous Production. *Journal of Pharmaceutical Innovation*, *10*(3), 191–199 (2015).
4. Pharmaceutical CGMPs for the 21s Century - A risk-based approach. Food Drug Adm. 32 (2004).
5. Chew, C. M., Ristic, R. I., Dennehy, R. D., & De Yoreo, J. J. Crystallization of paracetamol under oscillatory flow mixing conditions. Cryst. Growth Des., 4(5), 1045-1052 (2004).
6. Ristic, R. I. Oscillatory mixing for crystallization of high crystal perfection pharmaceuticals. Chem. Eng. Res. Dev., 85(7), 937-944 (2007).
7. Lindenberg, C., Krattli, M., Cornel, J., Mazzotti, M. Design and Optimization of a Combined Cooling/Antisolvent Crystallization Process. Cryst. Growth Des. 9(2), 1124–1136 (2009).
8. Global Burden of Disease Collaborative Network. Global Burden of Disease Study 2016 Results. Seattle, United States: Institute for Health Metrics and Evaluation (IHME) (2017).

Publication no.: IPDOIPM-00657

627822 Development of a Novel Continuous Spatially Distributed Diafiltration Unit Operation

Abstract:

Flow chemical and continuous processes have seen significant uptake in R&D and small intensified manufacturing for complex chemical and pharmaceutical products. However, continuous work-up and purification operations for continuous manufacturing requires further development to approach the significant level of purification and stream adjustment in multistep routes that batch synthesis traditionally offers in the production of these classes of chemical products.

The objective of this study is to develop an operation that can conduct separations based on diafiltration using semipermeable nanofiltration or ultrafiltration membranes in a fully continuous manner in a single stage configuration. To this end a Continuous Spatially Distributed Diafiltration (CS2D) operation was developed, to conduct straight through diafiltration in a manner that would yield equivalent or better purification efficiency, yield and productivity to batch diafiltration. To achieve this goal diafiltration solvent is introduced spatially in a highly uniform manner with the flow guided by a range of 3D-printed mixers designed using CFD informed design to increase localized mixing of retentate and diavolumes flows or displacement effects within the channel. Static mixers were 3D-printed using titanium and polyether ether ketone (PEEK) providing high pressure and chemical compatibility (Figure 1), suitable for many flow chemical separations.

Figure 1: (a) mixer filled membrane flatform (1-bottom plate; 2- middle plate to hold membrane with double O-ring; 3-mixer; 4-top plate) and (b) selection of printed mixers

Ibuprofen was selected as a model active pharmaceutical ingredient (API) with methanol acting as a solvent and ethanol used as a model impurity where an organic solvent nanofiltration membrane was used to selectively retain ibuprofen. Experiments were processed at a 55 bar transmembrane pressure at which 87% and 15% rejection coefficients of ibuprofen and ethanol were determined respectively, in screening experiments. Using 4 diavolumes in constant volume diafiltration in batch and CS2D runs, purities of 55% and 76% were obtained at equivalent yields of approximately 61% for the Batch and CS2D separation respectively. Demonstrating that continuous spatially distributed diafiltration (CS2D) can match or exceed results conducted in equivalent batch configurations. As such CS2D may present an attractive intermediate purification and solvent swap operation for telescoped flow chemical and continuous processing applications, in addition to a highly compatible platform for use in automated flow-based experimentation.

Acknowledgements

This publication is supported SSPC, the Science Foundation Ireland Research Centre for Pharmaceuticals (12/RC/2275_P2) and I-form the Science Foundation Ireland Centre For Advanced manufacturing (16/RC/3872) and TIDA, SFI/EI Technology Innovation Development award (16/TIDA/4003) co-funded under the European Regional Development Fund.

(a)

(b)

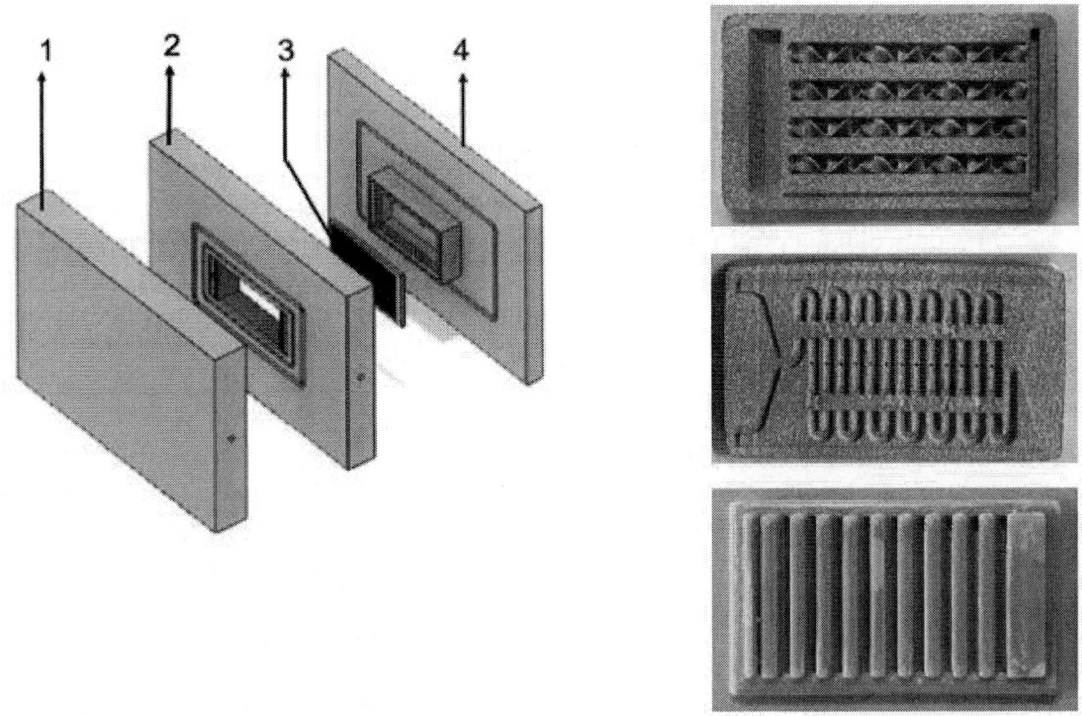

624587 Predicting Screw Feeder Flow Rates from Powder Properties and Operating Conditions

Abstract:

A well-known bottleneck in drug product development is the dependence on experimentally fit semi-empirical flowsheet models to design continuous manufacturing processes. The creation of predictive flowsheet models that account for powder properties and operating conditions effects on powder flow loosens this bottleneck by enabling engineers to begin design upon characterization and inform experimentation, ultimately speeding up a product's time-to-market. Such processes begin with identifying several screw feeders which are responsible for controlling the mass flow rates of active pharmaceutical ingredients and excipients across the entire process. Prior work has shown screw feeder mass flow rates are related to both the feeder configuration [1]–[3] and a powder's characterized properties [4]–[8], but do not provide a framework for predictively simulating a screw feeder's dynamics. Our paper addresses this process development gap by proposing a predictive, high-fidelity flowsheet model that can provide insight into how powder characteristics, feeder configuration, and operating state influence a screw feeder's dynamic behavior.

Our predictive feeder model is built from three components: a simple volumetric displacement screw feeder flowsheet model, a statistical autoregressive moving average model (ARMA), and a predictive projection to latent spaces (PLS) deterministic model. First, the hybrid deterministic displacement and stochastic ARMA model, as proposed by Johnson et al. [9], were fit to experimental runs of five powders with varying property values and feeding behaviors using both Coperion K-tron KT20 and MT12 feeders with twin concave screws operated at many different volumetric and gravimetric setpoints. This yielded sets of deterministic and stochastic deviation model parameters for each powder-feeder-setpoint operating configuration. Next, a PLS model was built that maps a database of powder properties [10], an abstraction of feeder geometries, and operating state to their experimentally estimated deterministic and stochastic parameters, extending the work of Bostijn et al. [8]. Constraints were integrated into the PLS model as needed to ensure deterministic parameters are physically possible and stochastic ARMA parameters meet theoretical requirements for stability and invertibility. The PLS model scores and loadings are presented, providing insight into how different powder properties influence the minute-by-minute and second-by-second dynamic feeding behavior. Last, the predictive capabilities of our PLS model are demonstrated and evaluated by simulating the feeding behavior of out of sample powders and conditions.

[1] H. Berthiaux, K. Marikh, and C. Gatumel, "Continuous mixing of powder mixtures with pharmaceutical process constraints," *Chemical Engineering and Processing: Process Intensification*, vol. 47, pp. 2315–2322, 2008.

[2] J. W. Fernandez, P. W. Cleary, and W. McBride, "Effect of screw design on hopper drawdown of spherical particles in a horizontal screw feeder," *Chemical Engineering Science*, vol. 66, pp. 5585–5601, 2011.

[3] W. E. Engisch and F. J. Muzzio, "Method for characterization of loss-in-weight feeder equipment," *Powder Technology*, vol. 228, pp. 395–403, 2012.

[4] L. X. Liu, I. Marziano, A. C. Bentham, J. D. Litster, E. T. White, and T. Howes, "Effect of particle properties on the flowability of ibuprofen powders," *International journal of pharmaceutics*, vol. 362,

pp. 109–117, 2008.

[5] M. Capece, K. R. Silva, D. Sunkara, J. Strong, and P. Gao, "On the relationship of inter-particle cohesiveness and bulk powder behavior: Flowability of pharmaceutical powders," *International Journal of Pharmaceutics*, vol. 511, pp. 178–189, 2016.

[6] Y. Wang, T. Li, F. J. Muzzio, and B. J. Glasser, "Predicting feeder performance based on material flow properties," *Powder Technology*, vol. 308, pp. 135–148, 2017.

[7] U. V. Shah, V. Karde, C. Ghoroi, and J. Y. Y. Heng, "Influence of particle properties on powder bulk behaviour and processability," *International journal of pharmaceutics*, vol. 518, pp. 138–154, 2017.

[8] N. Bostijn, J. Dhondt, A. Ryckaert, E. Szabó, W. Dhondt, B. Van Snick, V. Vanhoorne, Ch. Vervaet, and Th. De Beer, "A multivariate approach to predict the volumetric and gravimetric feeding behavior of a low feed rate feeder based on raw material properties," *International Journal of Pharmaceutics*, vol. 557, pp. 342–353, 2019.

[9] B. J. Johnson, S. García-Muñoz, M. Sen, J. Hanson, D. Slade, and N. V. Sahinidis, "(366d) stochastic analysis and modeling of pharmaceutical screw feeder mass flow rates," in *Proceedings of the 2020 Virtual AIChE Annual Meeting*, 2020.

[10] B. Van Snick, J. Dhondt, K. Pandelaere, J. Bertels, R. Mertens, D. Klingeleers, G. Di Pretoro, J. P. Remon, Ch. Vervaet, Th. De Beer, V. Vanhoorne, "A multivariate raw material property database to facilitate drug product development and enable in-silico design of pharmaceutical dry powder processes," *International journal of pharmaceutics*, vol. 549, pp. 415–435, 2018.

631153 Model-Based Analysis of Breakage in Fluid Bed Drying of Continuously Produced Pharmaceutical Wet Granules

Abstract:

In continuous pharmaceutical wet granulation, subsequent drying is a necessary step for evaporating the granulation liquid from the granules and rendering these processable downstream and in tableting. In the ConsiGma™-25 wet granulation this is achieved through semicontinuous fluid bed drying. The desired effect of drying to the ideal moisture content is accompanied by breakage of granules during this process (De Leersnyder et al., 2018). The generated fines have a negative impact on the processability downstream, such as possibly reduced content uniformity and flowability, as well as to the drying stage itself where the fine material gets stuck on the filter that entrains particulate material in the outgoing air. The latter could affect the duration limit of continuous operation as the pressure across the filters increases due to the fouling.

Not only does this breakage cause lower processability, it is also complex to capture and predict. It namely occurs in different locations in the process, within each of a different nature. Wet granules exiting the granulator are sent to the dryer through gravitational loading in a vertical ConsiGma™ setup, whereas in horizontal configurations these are pneumatically conveyed through a tube. Through contact with the tube wall granules undergo size reduction, dependent on their current properties governed by composition, size and moisture content. The liquid moreover provides the granules with different elastic properties than in a dry state. The next stage is during the fluid bed drying itself, as granules are dropped into the fluidizing air from the top of the dryer and collide with the dryer cell walls or already present granules. This breakage extent is affected by dryer process settings and granule properties. Finally, often the largest size reduction of the product occurs in the discharge of the dried granules to the subsequent evaluation module. A more forceful pneumatic conveying transports the granules through another tube, termed dry transfer line, causing breakage in function of granule properties, most prominently remaining moisture content.

As this problem is too complex to forecast, mathematical models can be used to predict the extent and nature of breakage occurring due to location, material and granule properties. A change in particle size distribution (PSD) can be modeled by means of a population balance model (PBM). Breakage models can be implemented in this framework, with kernels describing (combinations of) different fragmentation mechanisms that are plausible for the system at hand. The extent of breakage and other parameters of these functions can be mathematically linked to the modeled materials' properties by using PSD data.
The quality of such developed models is however not only dependent on the supplied data, also applying the right objective function plays a key role in finding the model and mechanism that describes the system in the most robust way. PSD data contains an abundance of information, it is therefore difficult to find an objective function that can take all of it into account. Also the experimental variability of PSD results could be used in model calibration in order to emphasize the most certain experimental information in the distributions. This work therefore implements techniques from the compositional data field to calculate variances on experimental PSD outcomes in function of size class. The (inverse of) the measurement uncertainty is then taken into account in the objective function. It has been shown that this allows to include information in both number- and volume-based PSD for mechanistic purposes (Ghijs, 2020).

Recent data collections have been carried out on the ConsiGma™-25 that varied the active pharmaceutical ingredient (API) and its powder properties with a fixed combination of excipients of 30% lactose monohydrate (Pharmatose 200M, DFE Pharma, Veghel, The Netherlands), 15% MCC (Avicel® PH-101, FMC International Health and Nutrition Cork, Ireland) and 5% HPMC (Methocel™ E15, The Dow Chemical Company, Midland, U.S.A.). The applied APIs showed different degrees of hydrophilicity and solubility. For each API, granules were produced at different levels of liquid addition. Part of the study also involved producing granules of similar properties based on different APIs, in order to isolate the effect of the applied API. Drying experiments were then carried varying the dryer inlet air temperature, mass loading and drying time. The experimental outcomes were characterized by high resolution PSD measurements (dynamic image analysis, QICPIC, Sympatec Gmbh, Clausthal-Zellerfeld, Germany) of the incoming and outgoing granules of the dryer.

In combination with the breakage PBM including measurement uncertainty in model calibration and validation, this study elucidates the effect of the tested API properties, granule properties (granule size and moisture content) and dryer process settings on their breakage behavior through the ConsiGma™-25 system. The more robust calibration helps in picking out the most appropriate breakage kinetics out of the abundance of PSD data. Moreover, a predictive model is generated that can take process settings, material and granule properties into account in forecasting breakage over the dryer for the selected formulations. This constitutes a pivotal modelling step towards a formulation-generic breakage model.

References

De Leersnyder, F., Vanhoorne, V., Bekaert, H., Vercruysse, J., Ghijs, M., Bostijn, N., Verstraeten, M., Cappuyns, P., Van Assche, I., Vander Heyden, Y., Ziemons, E., Remon, J. P., Nopens, I., Vervaet, C., & De Beer, T. (2018). Breakage and drying behaviour of granules in a continuous fluid bed dryer: Influence of process parameters and wet granule transfer. *European Journal of Pharmaceutical Sciences*, *115*, 223–232. https://doi.org/10.1016/j.ejps.2018.01.037

Ghijs, M. (2020). *Modelling and experimentation on fluid bed drying of granules in the context of pharmaceutical continuous wet granulation*. Universiteit Gent. Faculteit Bio-ingenieurswetenschappen.

629342 Development of a DEM-Based Digital Twin of a Continuous Direct Compression Line

Abstract:

A Continuous Direct Compression line (CDC) is a minimal setup for the production of solid dosage forms and consists only of three unit operations: feeding of the individual components, mixing the individual material streams, and compressing the blend into tablets. The discrete element method (DEM) is a powerful tool for in-depth investigations of the individual unit operations, however, the computational effort does not allow to model a complete CDC line at once.

In order to create digital twin of the CDC line, reduced order models are needed. A popular implementation of the reduced order models is based on residence time distributions (RTDs) coming from tracer experiments [1] or simulations [2]. The digital twin is able to predict the content uniformity after each processing step and in the final tablets with real feeder fluctuation data and under constructed disturbance scenarios.

In addition to exemplary scenarios, the digital twin is able to create so-called funnel plots. The funnel plots show the maximum duration and intensity of disturbances that still allow the production of tablets with in-spec content uniformity without any control action. The digital twin is thus a valuable tool to develop the process control strategy for the CDC line.

References

[1] Lee et al., 2021. " Continuous Mixing Technology: Characterization of a vertical mixer using residence time distribution". Int J Pharm, in press. https://doi.org/10.1016/j.xphs.2021.01.035

[2] Toson et al. "Continuous Powder Mixing Technology: Validation of the DEM Model". submitted to J Pharm Sci.

624558 Physical Property Characterization of Ionic Liquids in HFC-32 and HFC-125 Using a High-Pressure View Cell

Abstract:

The current hydrofluorocarbon (HFC) refrigerants used in automobile air conditioning, home appliances and air-conditioners, and retail refrigerant systems are being phased out due to high global warming potentials, resulting in millions of kilograms of refrigerant needing to be recycled and repurposed. Many of these refrigerants are HFC mixtures, and if separated, the components may be repurposed or reused to prevent release of greenhouse gases into the atmosphere. Since most HFC mixtures are azeotropic, separation is difficult and energy intensive and thus no reasonable technologies exist today. Ionic liquids (ILs) offer a sustainable way to separate HFC refrigerants if an IL can be found with a large solubility difference between the components of an HFC mixture. This research is part of Project EARTH (Environmentally Applied Research Toward Hydrofluorocarbons) that is working to create an efficient and sustainable process for the separation of HFC refrigerants. The group has measured the solubility of HFC-125 (CHF_2CF_3) and HFC-32 (CH_2F_2), which make up the refrigerant R-410A (50-50 wt%), in selected ILs using a microbalance. The microbalance does not allow for visually observing what phase the IL might exist in when pressurized with the refrigerant, therefore this project utilizes a high-pressure view cell with quartz windows to visually study the solid-liquid phase change of imidazolium halide based ILs. In some cases, the ILs are solid at room temperature, but in the presence of HFC-125 and HFC-32 as a function of pressure melt and undergo a phase change. Understanding the phase behavior of an IL during HFC sorption allows for the management of heat in separation processes as heat of fusion can help reduce the heat of sorption. Other property measurements based on visual data, including melting point, the presence of metastable states, and occurrence of foaming, are taken in addition to phase change data. This presentation will be discussing how the physical properties of imidazolium halide based ILs change with the sorption of HFC-32 and HFC-125 to indicate the viability of the ILs for separation of the refrigerant R-410A.

629168 Material Selection and Design of Ionic Liquid-Based Extractive Distillation for Hydrofluorocarbons Separation

Abstract:

Hydrofluorocarbons (HFCs) and their mixtures are used as refrigerants for commercial and domestic cooling systems. Because of high global warming potential (GWP) of some of these mixtures, the Kyoto Protocol has recommended their gradual phase-out. There is an economic incentive to separate these HFC mixtures into their constituent components rather than venting or incinerating. However, many of these HFC mixtures are azeotropic in nature, and there is no commercial technology available to perform the separation. Extractive distillation, first proposed in the 1930s [1], is a technology that can separate close boiling azeotropic mixtures by using a higher boiling point entrainer. Ionic liquids (ILs) are attracting attention as entrainer for the extractive distillation of HFCs because of their negligible vapor pressure and chemical and thermal stability [2-3]. Due to significantly large number of possible ions combinations, theoretically, it is possible to have millions of ILs [4]. Selecting an appropriate IL for the separation of a particular HFC mixture is important for effective and economic distillation [5]. We have developed a multiscale process design, synthesis, and material selection framework called SPICE_ED for the synthesis of optimal extractive distillation process for separating HFC mixtures. SPICE_ED stands for Synthesis and Process Intensification of Chemical Enterprises involving Extractive Distillation. The framework is based on a building block-based novel representation of chemical phenomena, tasks, and operations in the chemical process industry [6-8]. From a candidate set, the framework can automatically select the most optimal IL that offers minimum energy requirement and/or has minimum processing cost. We generate process flowsheets and equipment designs considering the appropriate refrigerant and IL inlet trays, IL flow rates, reflux ratio, column operating conditions, and IL regeneration processes involving separate distillation, stripping, and flash operations. Finally, the framework can be utilized to map out the thermophysical properties of hypothetical ILs, thus directing the efforts towards discovering novel ILs with desired process economics. When searching for an optimal process for the separation of R-410A (50 wt% R-125 (C_2HF_5) and 50 wt% R-32 (CH_2F_2)), we obtained a heat-integrated optimized process that consumes 65% less energy and emits 72% less indirect CO_2 emission compared to previously reported designs with an existing IL [9]. In this presentation, we will also discuss how the framework can be employed for large-scale computational screening of ILs and other solvents for the separation of mixers with optimized process designs.

References:

[1] Randall, M. and Webb, W.A., 1939. Separation process. *Industrial & Engineering Chemistry, 31*(2), pp.227-230.

[2] Morais, A.R.C., Harders, A.N., Baca, K.R., Olsen, G.M., Befort, B.J., Dowling, A.W., Maginn, E.J. and Shiflett, M.B., 2020. Phase Equilibria, Diffusivities, and Equation of State Modeling of HFC-32 and HFC-125 in Imidazolium-Based Ionic Liquids for the Separation of R-410A. *Industrial & Engineering Chemistry Research, 59*(40), pp.18222-18235.

[3] Shiflett, M.B. and Maginn, E.J., 2017. The solubility of gases in ionic liquids. *AIChE Journal, 63*(11), pp. 4722-4737.

[4] Shiflett, M.B. and Yokozeki, A., 2006. Solubility and diffusivity of hydrofluorocarbons in

room-temperature ionic liquids. *AIChE Journal*, 52(3), pp.1205-1219.

[5] Lei, Z., Dai, C., Zhu, J. and Chen, B., 2014. Extractive distillation with ionic liquids: a review. *AIChE Journal*, *60*(9), pp.3312-3329.

[6] Demirel, S.E., Li, J. and Hasan, M.M.F., 2017. Systematic process intensification using building blocks. *Computers & Chemical Engineering*, *105*, pp.2-38.

[7] Demirel, S.E., Li, J. and Hasan, M.M.F., 2019. A General Framework for Process Synthesis, Integration, and Intensification. *Industrial & Engineering Chemistry Research*, *58*(15), pp.5950–5967.

[8] Demirel, S.E., Li, J., El-Halwagi, M.M. and Hasan, M.M.F., 2020. Sustainable Process Intensification using Building Blocks. *ACS Sustainable Chemistry & Engineering*, *8*(48), pp.17664–17679.

[9] Shiflett, M.B. and Yokozeki, A., 2006. Separation of difluoromethane and pentafluoroethane by extractive distillation using ionic liquid. *Chimica oggi*, *24*(2), pp.28-30.

631872 The Effect of Alkyl Chain Length on Imidazolium-Based Ionic Liquids and Choice of Halide Anions (Cl, Br, I) on Phase Equilibria for HFC-32, HFC-125, and Binary Mixtures (R-410A)

Abstract:

The current generation of refrigerants, hydrofluorocarbons (HFCs), are being phased out due to their high global warming potential (GWP) while the next generation of more environmentally friendly refrigerants, hydrofluoroolefins (HFOs), are developed and commercialized. Consequently, millions of tons of HFC mixtures will need to be recycled. Project EARTH (Environmentally Applied Research Toward Hydrofluorocarbons) is focused on developing efficient and selective separation processes for HFC mixtures. One promising approach is to use ionic liquids (ILs) as an entrainer to selectively separate azeotropic HFC mixtures like R-410A, which is composed of 50 wt.% HFC-125 (CHF_2CF_3) and 50 wt.% HFC-32 (CH_2F_2). The HFC-32 can be blended with HFOs to produce new products such as R-454B (68.9 wt.% HFC-32 and 31.1 wt.% HFO-1234yf ($CH_2{=}CFCF_3$)) and HFC-125 can be repurposed as a high-value fluorinated feedstock for production of new materials with a low-GWP.

This presentation will compare the effects of alkyl chain length on imidazolium-based ILs and anion substitution (Cl, Br, I) on the phase equilibria data for HFC-32 and HFC-125 and mixtures (e.g., R-410A). This presentation will also discuss ILs that are both solid and liquid at 25 °C. Phase equilibria data for HFC-32 and HFC-125 were measured using a Hiden Isochema IGA gravimetric balance and mixtures of HFC-32 and HFC-125 were measured using the first Hiden Isochema XEMIS Integral Mass Balance (IMB) method. The impact of alkyl chain length and anion on diffusion and selectivity will also be discussed for both pure HFC+IL and binary HFC+IL mixture measurements.

631738 Liquid-Liquid Equilibria of Ternary Systems of Thermally Robust Ionic Liquids and Hydrocarbon Mixtures

Abstract:

Hydrocarbon (HC) separations is an energy intensive process due to similar boiling points and formation of azeotropes. Here, we test different novel thermally robust perarylphosphonium and perarylsulfonium ionic liquids (ILs) for separating aromatic compounds from HC mixtures. Molecular dynamics (MD) simulations are used to provide a molecular-level understanding of these systems. Simulations of binary mixtures of toluene/ILs and *n*-heptane/ILs are performed to validate and study the miscibility behavior. Ternary mixtures consisting of IL/toluene/*n*-heptane at different concentrations of toluene in *n*-heptane are carried out to determine the amount of toluene absorbed into IL phase. Further, the fundamental interactions between aromatics and IL leading to this behavior are assessed through radial distribution functions and nearest neighbor analysis. Selectivity and distribution coefficients are determined to evaluate the affinity of toluene between IL and HC phase. The free energy of solvation ($\Delta G_{solvation}$) and transfer free energy ($\Delta G_{transfer}$) of toluene in different solvents are calculated using thermodynamic integration (TI) method. The results show that $\Delta G_{solvation}$ of toluene is more favorable in all the ILs compared to *n*-heptane. Overall the simulation results provide a molecular-level understanding of the solvent environment provided by these ILs in separating aromatics from hydrocarbon mixtures.

629325 The Deterministic Role of Hydrogen Bonding on Water Diffusivity in Methylimidazolium Halide Ionic Liquids and Ionogels

Abstract:

Polymeric ionogels, crosslinked gels swollen by ionic liquids (ILs), are useful vehicles for the release and storage of molecular solutes in next generation separation and delivery applications. Although rapid solute diffusion is often critical for performance, it remains challenging to predict diffusivities across multi-dimensional composition spaces. Recently, we showed that water (a model neutral solute and ubiquitous contaminant) diffuses through alkyl-methylimidazolium halide ILs by hopping between hydrogen bonding sites on relatively immobile cations. Here, we expand on this activated hopping mechanism in two significant ways. First, we demonstrate that water diffuses through poly(ethylene glycol) diacrylate ionogels via the same mechanism at a reduced rate. Second, we hypothesize that the activation energy barrier can be determined from relatively simple ^{1}H NMR chemical shift measurements of the proton responsible for H-bonding. This relationship enables water's diffusivity in ionogels of this class to be predicted quantitatively, requiring only (1) the composition-dependent diffusivity and Arrhenius behavior of a single IL, and (2) ^{1}H NMR spectra of the ionogels of interest. High-throughput microfluidic Fabry-Perot interferometry measurements verify prediction accuracy across a broad formulation space (4 ILs, $0 \leq x_{H2O} \leq 0.7$, $0 \leq \varphi_{PEGDA} \leq 0.66$). The predictive model may expedite IL material screening. Moreover, it intimates a powerful connection between solute mobility and hydrogen bonding, and suggests targets for rational design.

624169 Applications of Phytoremediation for Catalytic Enhancement of Biocarbon Towards Supercapacitors

Abstract:

The U.S. generates approximately two billion tons of mining waste annually, leading to inorganic heavy metal contaminates such as Ni, deposited into surrounding ecosystems at toxic levels. Phytoremediation offers a low-cost solution for environmental clean-up, particularly in impoverished areas, however current phytoremediation strategies generate unusable biomass due to the toxic metal contents. A novel solution for this unused waste lies in the area of energy storage, especially for supercapacitor applications. This work focuses on the cultivation of water hyacinths, a known metal hyperaccumulator, in various concentrations of Ni-doped water followed by synthesis of harvested biomass into high quality biocarbon. Biomass was subjected to thermal annealing at 500 °C for 1 h under Argon environment to cleave functional groups. Subsequent biochar was improved through thermochemical/catalytic activation at 800 °C for 1 h with KOH providing additional enhancement during biocarbon synthesis. Electrochemical results demonstrate a specific capacitance of 541 F g^{-1} making it suitable for use in the supercapacitor for electricity storage. Physical and electrochemical characterizations, including ICP, SEM, BET, ATR-FTIR, Raman, and XRD, cyclic voltammetry, chronopotentiometry, and EIS will be discussed. This work provides a value-added product and in-depth insights into naturally embedded catalyst, while simultaneously remediating contaminated mining sites.

624185 Electrochemical Studies on Graphitized Biocarbon Derived from Hydrothermally Liquefied Low Ash Content Corn Stover

Abstract:

Biomass-derived biocarbon has gained considerable attention due to its abundance, low-cost, and chemical structures conducive for energy storage applications, particularly supercapacitors. Lignocellulosic components embedded within the cell wall of biomass substrates can provide a fine template for enhanced ion storage, transport, and rate capabilities, desirable for electrochemical storage. Due to its abundant availability of agricultural residues such as Corn Stover, in this research we have utilized the homogenized low ash content corn stover milled and sieved to desired specifications. As obtained corn stover is converted to hydrochar via catalytic hydrothermal liquefaction in the presence of $Ni(NO_3)_2$ catalyst at 275°C and its morphological properties such as surface area, pore volume and diameter were enhanced in the presence of $ZnCl_2$, followed by thermal activation at 450°C. Later as derived hydrochar is pH neutralized and acid washed with dilute HCl to extract metal contaminants. Biochar underwent thermal carbonization at 850 °C for 3 h to enhance pore structures and graphitization, conducive for electrochemical applications. Electrochemical results demonstrated specific capacitances in excess of 300 F g^{-1}, comparable to literature values for corn stover-derived biocarbons. Physical and electrochemical characterizations, such as BET, ATR-FTIR, Raman, XRD, cyclic voltammetry and chronopotentiometry, will be discussed to give fundamental insights into biocarbon enhancement. The results obtained during the electrochemical characterization and studies will be presented.

625756 Electrocatalyst Development for Making Ammonia from Air and Water at Ambient Condition

Abstract:

Renewable power generated from wave energy has faced technological and cost barriers that have hindered its penetration into utility-scale electricity markets. As an alternative, the production of chemical fuels—for example, ammonia (NH_3), which has a high energy density (11.5 MJ/L) and facile storage properties—may open wave energy to new markets including ocean exploration and transportation. We developed a unique catalyst and tested its performance toward nitrogen reduction reaction using the rotating round disk electrode (RRDE) method. The current density increased 0.25 mA/cm^2 after switching the gas from Ar to N_2 and the production rate of NH_3 and Faradaic efficiency (FE) are in the top list of reported results.

627567 Ammonium Formate As a Safe, Energy-Dense Electrochemical Fuel Ionic Liquid

Abstract:

Hydrogen has been investigated widely as an energy carrier for clean electrons, however its low energy density and large energy losses incurred in compression have limited its use. For these reasons, solid and liquid carriers of hydrogen fuel have been investigated, including liquid ammonia, formic acid/formate salts, and methanol, to name a few. While these proposed fuels have significant potential, many of them do not meet the energy efficiency requirements necessary to outperform compressed hydrogen and enable facile transport and storage of renewable energy. For example, ammonia is an appealing energy carrier because of its high hydrogen content: 110 kg H_2/m^3 for liquid ammonia compared to 40 kg H_2/m^3 for compressed hydrogen at 700 bar and 288 K. It is already produced, transported, and stored at scale, and there has been recent progress in reducing the carbon footprint of ammonia production technologies. Formic acid is also appealing due to its hydrogen content (53 kg H_2/m^3), its ease of transport and storage since it is a liquid, and its potential to be synthesized efficiently from carbon dioxide, water, and renewable electricity. However, both ammonia and formic acid suffer from safety issues because they are caustic and corrosive, as well as difficulties with efficient release of energy at scale.

In this work, we propose ammonium formate, a combination of ammonia and formic acid, as an energy carrier. Not only is its production as cheap and simple as its constituent molecules, ammonium formate is a benign solid at ambient conditions, making it safe to transport and store. We demonstrate an electrochemical cell that takes advantage of the fact that ammonium formate forms an intrinsically conductive ionic liquid at 116°C; as an ionic liquid, ammonium formate has negligible vapor pressure and is significantly safer than either ammonia or formic acid alone. In this system, hydrogen is evolved at the cathode with near 100% Faradaic efficiency and formate is oxidized to carbon dioxide at the anode with near 100% Faradaic efficiency. In theory, ammonia could be oxidized to nitrogen in the same cell, however we observe that the operating temperature makes it so that ammonia gas evaporates from the cell and can be oxidized using a second, modular ammonia fuel cell. Because molten ammonium formate is conductive, the electrolyte and feed can approach 100% ammonium formate fuel, resulting in a large thermodynamic driving force. Overall, this system represents a new class of electrochemical fuel ionic liquids and results in modular release of H_2 with the potential for zero net carbon emissions.

632405 Enhanced CO_2 Electroreduction to CH_4 and C_2H_4 Via Selective Proton Transfer

Abstract:

CO_2-derived fuels, synthesized using renewable energy, present an attractive route towards fuel formation from sustainable energy. The role of proton transfer in dictating the efficiency and selectivity of the two-electron, two-proton reduction of CO_2 to CO has been well-studied. Indeed, impressive progress has been achieved in the production of CO from CO_2. However, the role of proton transfer to surface-bound CO intermediates in governing the selective electrosynthesis of higher order products beyond CO, such as methane or ethylene, remains poorly understood.

Here, we investigate the electrochemical reduction of CO at Cu electrodes in nonaqueous electrolytes at low temperatures. The choice of the electrolyte and reaction conditions allows fine-tuned control of the proton donor and the CO binding strength, enabling for the first time activation-controlled kinetic studies over an extended parameter range. Our studies lead to the surprising finding that increasing the concentration of CO serves to suppress methane production, in addition to hydrogen. In contrast, ethylene production remains unaffected by CO concentration and depends predominantly on the electrochemical potential.

These observations provide unprecedented insight into the mechanism of electrochemical CO and CO_2 reduction. Among others, we are able to show that the rate of methane and hydrogen formation is governed by the competition of CO and H for surface sites and that ethylene formation is taking place from sites which are saturated with CO, indicating that simply increasing the supply of CO may not be a viable strategy for improving CO and CO_2 reduction systems. Based on our results, we subsequently demonstrate how knowledge of the reaction mechanism can be exploited to selectively suppress hydrogen evolution relative to CO reduction by rational tuning of the proton donor.

638245 Bridging the Knowledge Gap by Empowering Engineers with IIoT Advanced Analytics

Abstract:

You've been gathering production process data for years, but are you leveraging its full potential? Enabling your operational experts to turn data into actionable information is crucial in accelerating growth, controlling product quality and increasing sustainability.

Whether your goals are related to energy management, waste reduction, or increasing uptime, having your operational data at your fingertips is critical.

In this presentation, we will demonstrate how the practical use of self-service data analytics allows process engineers and operators to leverage all of their historical and time-series data - without having to rely on data scientists and spreadsheets.

635600 Biodirected Assembly of Ionomer Thin Films for Fuel Cells and Electrolyzers

Abstract:

Increasing evidence suggests that control of ionomer thin film structures on metal surfaces is pivotal for efficient performance in electrochemical devices such as fuel cells and electrolyzers. Unfortunately, the assembly of these thin film structures is difficult to control using conventional methods and ideal arrangements remain unknown. Engineered polypeptides have emerged as powerful biomolecular tools in electrode assembly because binding sites and polypeptide structures can be easily modulated by changing the amino acid sequence. However, no studies have been conducted showing polypeptides can be engineered to interact with ionomers, attach them to metal surfaces, and control their arrangement. Our lab has recently developed this technology, using an elastin-like polypeptide to bind to metals, and bind to acidic and basic ionomer via ionic interactions. We use a quartz crystal microbalance with dissipation to provide detailed information about the loading, thickness, and binding behavior of the polypeptide and ionomer layers. We also use atomic force microscopy and grazing-incidence small-angle X-ray scattering to understand the impact of the biomolecules on ionomer phase separation. Finally, we have begun to analyze the performance (ionic conductivity) of the assembled films using interdigitated electrodes and electrochemical impedance spectroscopy. Through these techniques, we show that 1) our biomolecular system is highly flexible, easily adapting to different materials, 2) the polypeptide sequence can dictate ionomer phase separated structures, and 3) this system can be used to improve the performance of ionomer thin films and gain structure-function understanding. Generally, our results demonstrate that engineered polypeptides are promising tools for ionomer control in electrode engineering.

635530 Parametric Testing, Analysis and Techno-Economic Assessment of 2-Eempa, a Single-Component Water-Lean Post-Combustion Capture Solvent

Abstract:

We present here, the design, comprehensive testing and technoeconomic assessment of N-(2-ethoxyethyl)-3-morpholinopropan-1-amine (2-EEMPA), a single-component water-lean solvent developed at PNNL. This presentation details the physical and thermodynamic properties of the solvent, including vapor–liquid equilibria, kinetics and viscosity measurements in addition to over 40 hours of continuous CO_2 capture performance in a laboratory scale system on simulated flue gas comprised of expected levels of H_2O, O_2, NOx and SOx. A full techno economic analysis of the capture process at industrial scale with corresponding projections of critical metrics was performed using Aspen Plus. Here, we present that 2-EEMPA can be optimized for efficiency, with a projected regeneration heat rate of 2.0 GJ per tonne CO_2 for post-combustion capture of coal-derived flue gas. Conversely, we detail how 2-EEMPA can be optimized for cost using a 2-stage flash regeneration, where projected total cost of capture of $37.6/tonne (in 2018 USD) CO_2 using the US Department of Energy's revised cost and performance assumptions. We close with a discussion on further cost reduction potential and preparation for upcoming slip-stream testing of this solvent at a 0.5 MW scale.

635542 Diversification of Basic Immobilized Amine Sorbents (BIAS) for Carbon Dioxide Capture

Abstract:

Carbon sequestration entails a multi-step process in which anthropogenic CO_2 emissions are captured from CO_2-laden process gas streams and permanently stored. Carbon capture is a critical step in the process and accounts for a considerable portion of the overall cost. Across the literature, a multitude of functionalized silica sorbents demonstrate high-capacity CO_2 capture and offer advantages over the benchmark liquid monoethanolamine (MEA) process, including reduced heat duty and improved CO_2 mass transfer. However, low sorbent stability still remains an issue. Basic immobilized amine sorbents (BIAS) developed at NETL were successfully tested at bench and pilot-scales, and demonstrated their promising potential for large scale CO_2 removal from post-combustion sources and ambient air. This is owed to the BIAS' optimized tradeoff between CO_2 capture capacity and thermal stability observed over multiple and even 100 adsorption-desorption cycles. Within the last few years, key technological milestones diversifying the commercialization potential of the BIAS were achieved. These include in house (bench scale, 10 g) and commercial (3 kg) pelletization of particles into rods and spheres for moving bed rectors; incorporation into a non-aqueous particle/silicone oil slurry regenerated by microwave irradiation for replacing/retro-fitting the MEA process; and spinning into hollow chemisorption fiber sorbents (CHEFS) for rapid temperature swing adsorption systems. The proven scalability of BIAS combined with the diverse portfolio of material architecture makes this technology ideal for industrial CO_2 capture applications, including for coal combustion and gasification power generating systems; landfill gas cleanup, enhanced oil recovery (EOR); natural gas sweetening, and life support systems.

636022 CO_2 Capture Modeling, Energy Savings, and Heat Pump Integration

Abstract:

While amine-based CO_2 capture systems have been designed since 1927, concern over anthropomorphic climate change sparked renewed interest in the 1980's. Subsequent research has led to significant advances such as well-fit thermodynamic models, rigorous rate-based distillation simulations, and improved process designs. The aims of this presentation are to (1) summarize proposed process improvements; (2) demonstrate the strong and complex interactions of energy-saving schemes; (3) report an optimal process design integrating all of these improvements in Aspen Plus, and then evolutionally simplify the design through simulation results. In particular, we propose an optimum energy-saving design (U.S. Patent No. 10,751,664) that uses an absorption-driven heat pump together with a distributed cross heat exchanger and a stripper vapor condensate rerouting in order to reduce both the cooling and heating utility consumptions. The resulting predicted solvent regeneration energy for the integrated system is 1.67 GJ/tonne CO2 captured. China Petroleum and Chemical Corporation (SINOPEC) is currently implementing the process to recover 1 million tons per year of CO_2 from the power plant flue gas in the Shengli Oil Field, Shangdong Province, for tertiary oil recovery.

624825 Process Intensification for Direct Conversion of Biomass-Based Syngas to High Octane Gasoline

Abstract:

The conversion of lignocellulosic-derived carbon sources via the creation of a CO_2 rich syngas stream provides means to produce liquid hydrocarbon fuels, with advancements intended to create a cost-competitive method of sequential conversion via methanol and dimethyl ether intermediates. Previously developed methods with Cu/Beta zeolites enabled the conversion of dimethyl ether to linear and branched hydrocarbons, with high selectivities towards products with high-octane fuel properties.[1] Through process intensification, use of a single reactor to convert syngas to methanol and DME, then further to hydrocarbons, allows for reduced capitol and operation cost for the same chemistries that typically use 3 reactors instead of one. Utilization of commercially available methanol synthesis catalyst (Megamax 800, Clariant, CZA) and methanol dehydration catalyst (g-Al_2O_3) with the Cu/Beta catalyst allowed for conversion into hydrocarbon products in this single reactor set up. Reactor configuration with CZA and alumina catalysts mixed with or stacked in a bed physically above the Cu/Beta catalyst provided a system for the direct conversion of the syngas to hydrocarbons with improved yields and selectivities with higher net C_1 conversions for stacked bed configurations. Through control of the process conditions, greater conversion of C1 oxygenate intermediates (i.e., methanol and dimethyl ether) was achieved with concomitant increase in selectivities towards gasoline and jet fuel range olefin and paraffin products. Lower hydrocarbon number products and less naphthenes were observed for the conversion of syngas compared to DME conversion on only the Cu/Beta system with prominent selectivities observed for C_4, C_5, and C_7 hydrocarbons.[2] The DME and methanol formation rates, with understanding of the equilibria for both processes, were determined as important factors to allow for improved performance of the Cu/Beta. Determination of the factors which allow for tuning of selectivities and yields created an intensified process which allows for a "market-responsive" biorefinery design, which can produce high octane gasoline or jet fuel range hydrocarbons to meet demands for a more sustainable route to liquid fuels.

References:

(1)Schaidle, J. A.; Ruddy, D. A.; Habas, S. E.; Pan, M.; Zhang, G.; Miller, J. T.; Hensley, J. E. Conversion of Dimethyl Ether to 2,2,3-Trimethylbutane over a Cu/BEA Catalyst: Role of Cu Sites in Hydrogen Incorporation. *ACS Catal.* **2015**, *5* (3). 1794–1803.

(2)Ruddy, D. A.; Hensley, J. E.; Nash, C. P.; Tan, E. C. D.; Christensen, E.; Farberow, C. A.; Baddour, F. G.; Van Allsburg, K. M.; Schaidle, J. A. Methanol to High-Octane Gasoline within a Market-Responsive Biorefinery Concept Enabled by Catalysis. *Nat. Catal.* **2019**, *2* (7), 632–640.

630926 Carbon Oxide-Thermic Oxidation Process for Synthesis of Porous Silicon-Carbon Composite Anodes for Lithium-Ion Batteries

Abstract:

Carbon Oxide-Thermic Oxidation Process for Synthesis of Porous Silicon-Carbon Composite Anodes for Lithium-ion Batteries

Mohamed A. Elharati[a], Younghwan Cha[b], Qusay Bkour[a], Steven R. Saunders[a], M. Grant Norton[b], Min-Kyu Song[b,*], Su Ha[a,*]

[a]Voiland School of Chemical Engineering and Bioengineering, Washington State University, Pullman, WA 99164-2710, USA

[b]School of Mechanical and Materials Engineering, Washington State University, P.O. Box 642920, Pullman, WA 99164-2920, USA

Abstract

The silicon-based compounds have been investigated as a promising anode material for lithium-ion (Li-ion) batteries. In this work, we show that a novel carbon oxide thermic oxidation (CO_x-OP) process is not only a facile and low-cost technique to synthesize porous silicon-carbon (Si-C) composites for high-energy Li-ion batteries but also is the novel utilization technology of greenhouse gas. Furthermore, we demonstrate that the morphology/microstructure and performance of Si-C composites can be controlled by using different gas compositions (CO, CO_2, and H_2).

In-situ high-temperature X-ray diffraction analyses were performed to elucidate the phase transition during CO_x-OP. The pulsing experiments were conducted to illustrate the mechanism of the CO_x-OP process. The electron microscopy and X-ray microanalysis were carried out to investigate the effects of processing conditions on the morphology and microstructures of Si-C. The TGA and Raman analysis were performed to determine the amount and type of carbonaceous materials.

The synthesized porous Si-C via CO_x-OP process showed good electrochemical performance for Li-ion batteries, and their performance is strongly affected by the gas composition used in the CO_x-OP process. For example, the first lithiation and delithiation capacities of Si-C prepared with CO_2-rich gas stream were 2481 mAh/g and 2061 mAh/g, respectively, resulting in high coulombic efficiency of 83.1% . In contrast, the first lithiation and delithiation capacities were 2148 mAh/g and 1356 mAh/g, respectively, resulting in low coulombic efficiency of 63.2.% for Si-C prepared with CO-rich gas stream.

Keywords:

Porous silicon-carbon (Si-C) composites, carbon oxide thermic oxidation (CO_x-OP) process, lithium-ion battery.

622570 Optimal Design of Controlled Environment Agricultural Systems Under Market Uncertainty

Abstract:

Global population growth and increased climate variability have posed major challenges to traditional agricultural production [2,6]. As these trends persist, meeting increased consumer demand under heightened uncertainty will further stress the capabilities of our existing food production infrastructure [4,5]. For this reason, alternative methods are being sought to augment conventional food production. Controlled environment agriculture (CEA) is an alternative strategy that decouples crop production from environmental conditions through indoor cultivation. CEA offers a more sustainable alternative to conventional agriculture and presents opportunities for improved food security by increasing public access to nutritious fresh produce. Startup costs for CEA systems, however, are significantly higher than those of traditional growing operations, and therefore present a larger barrier to adoption. To accelerate the adoption of this potentially disruptive technology for next generation food production, a better understanding of the long-term economic viability of CEA technology is desired. To provide this valuable insight, we have developed the first economics-based optimal design and planning model for CEA systems.

To date, the majority of agricultural planning models have focused on traditional outdoor food production. Many of the existing models have been tailored to highly specific operations, few have employed mathematically rigorous optimization techniques, and only a handful have accounted for uncertainty, which is desired to realistically assess the long-term economic viability of the system [1,3,7]. Since existing models are not applicable to CEA systems, we have developed an optimal design and planning model that can be used to assess the long-term economic viability of a broad range of CEA systems. Our proposed model employs robust optimization to account for uncertainty in the design and planning stages of CEA production, which enables risk-averse investment decision making for long-term economic viability.

In this work, we present a novel methodology for the simultaneous robust design and planning of CEA systems under multi-period market price uncertainty/risk. This problem is cast as a bilevel program with a nonconvex outer program pertaining to the design objective and multiple semidefinite inner programs pertaining to mean-variance risk exposure over each period considered in the planning horizon. The bilevel program is reformulated as a nonconvex semi-infinite program with several semi-infinite constraints pertaining to the multi-period risk exposure over the horizon. The problem is solved to global optimality using a custom branch-and-bound strategy constructed with EAGO.jl [10] that exploits a combination of McCormick-based envelopes and Gurobi's nonconvex solver [8], to solve the relaxed (upper-bounding in the maximization sense) subproblems, and IPOPT [9] for local solution of the lower-bounding (in the maximization sense) subproblems. A solution to the model represents a design and operating schedule that is robust to worst-case market uncertainty and therefore provides a conservative basis for CEA engineering and investment decision-making.

Our proposed model has been formulated generally and can be used to optimize large-scale CEA systems for the cultivation of any crop portfolio via single or multi-mode cultivation. The optimization objective employed is to maximize the net present value of the CEA system over its expected lifespan. Constraints limiting risk exposure and production capacity are also included to ensure that the optimal design is both robust and practical. Capital and operating expenses, including those associated with system construction, cultivation equipment, resources, labor, and distribution, are accounted for in the economic objective through sub-models that capture the location, capacity, and crop allocation cost dependencies of the specific CEA system being modeled. These sub-models can be easily adjusted for different CEA configurations without structural changes to the main optimization algorithm. This not only enables efficient determination of the optimal design for a given CEA configuration, but also enables efficient comparison between multiple candidate systems for strategic adoption of CEA technology.

To evaluate the efficacy of our approach, the economic performance of the robust design was compared against that of naïve design and operating strategies through two case studies. In each of the presented case studies, an economically feasible design was identified, even when the naïve design was economically infeasible, and the improvement in system performance attributed to the implementation of a robust design was quantified. This approach represents, to our knowledge, the first robust optimization approach to CEA systems and is demonstrated to effectively increase the robustness of CEA systems to market uncertainty, improve the long-term economics of CEA systems over naïve operating strategies, and validate the economic viability of single and multi-mode CEA production of distinct crop portfolios. This work establishes a means of evaluating the economic viability of a wide array of CEA systems and serves as a baseline for future development of CEA design and planning models to accelerate the adoption of CEA technology.

References:

[1] Ahumada, O., Villalobos, J.R., 2009. Application of planning models in the agri-food supply chain: a review. Eur. J. Oper. Res. 196 (1), 1–20. doi: 10.1016/j.ejor.2008.02.014 .

[2] Deschênes, O., Greenstone, M., 2007. The economic impacts of climate change: evidence from agricultural output and random fluctuations in weather. Am. Econ. Rev. 97 (1), 354–385. doi: 10.1257/aer.97.1.354 .

[3] Dury, J., Schaller, N., Garcia, F., Reynaud, A., Bergez, J.E., 2012. Models to support cropping plan and crop rotation decisions. A review. Agron. Sustain. Dev. 32 (2), 567–580. doi: 10.1007/s13593- 011- 0037- x .

[4] Fedoroff, N.V., 2015. Food in a future of 10 billion. Agric. Food Secur. 4 (1), 11. doi: 10.1186/s40066- 015- 0031- 7 .

[5] Field, C.B., Barros, V., Stocker, T.F., Dahe, Q., 2012. Managing the Risks of Extreme 1083 Events and Disasters to Advance Climate Change adaptation: Special Report of the Intergovernmental Panel on Climate Change (IPCC). Cambridge University Press.

[6] Food and Agriculture Organization of the United Nations (FAO), 2018. The state of agricultural commodity markets 2018: Agricultural Trade, Climate Change and Food Security. Technical Report. FAO. URL http://www.fao.org/3/I9542EN/i9542en.pdf .

[7] Glen, J.J., 1987. Mathematical models in farm planning: a survey. Oper. Res. 35 (5), 641–666. doi: 10.1287/opre.35.5.641 .

[8] Gurobi Optimization, LLC, 2020. Gurobi optimizer reference manual. URL http://www.gurobi.com .

[9] Wächter, A., Biegler, L.T., 2006. On the implementation of an interior-point filter line-search algorithm for large-scale nonlinear programming. Math. Program. 106 (1), 25–57. doi:

10.1007/s10107- 004- 0559- y.

[10] Wilhelm, M.E., Stuber, M.D., 2020. EAGO.jl: Easy advanced global optimization in Julia. Optim. Methods Softw. doi: 10.1080/10556788.2020.1786566 .

632232 Reproducible and Fast Mapping of Material Flows in Industrial Networks Using Piot Hub: A Novel Cloud Based Computational Tool

Abstract:

An important goal of industrial ecology is to ensure a balance is maintained between economic and environmental sustainability. Facilitation of this balance requires an understanding of the structure of the material economy and the material flows/waste generation in industrial networks in order to minimize the environmental consequences of these industrial systems. Physical Input-Output Tables (PIOTs) are uniquely suited for this purpose by mapping physical flows between industrial sectors. In previous work, PIOT-Hub was developed to automate the generation of both Physical Supply Tables (PSTs) and Physical Use Tables (PUTs) as well as their conversion into PIOTs by integrating the strength of mechanistic models and macroeconomic Input-Output framework [1]. Hence, PIOT-Hub, a cloud-based tool can be used to generate the PST and PUT for a region using a bottom-up approach where relevant industrial processes are simulated via mechanistic models to determine steady-state material input and output requirements. Thus, the tool establishes a connection between engineering and economic model in such a way that it is insightful for transitioning to a low carbon and zero waste circular economy. In past work, PIOT-Hub was applied to 11 agro-based economic sectors in Illinois, thus capturing elemental flows across the economy, demonstrating the automation aspect. Here we evaluate the reproducibility and time aspect of PIOT-Hub to map the material flows of equivalent economic sectors in Indiana. We develop simulation models for industries in Indiana and utilize PIOT-Hub for mapping the industrial network, thus evaluating the functionality of this novel computational platform. This work not only demonstrates the reproducibility of the tool's utilization to other geographic regions, but also serves as a case study of how the fast generation of PSTs, PUTs, and PIOTs by PIOT-Hub can be used to compare the material economies of multiple regions. We perform a comparative analysis of material flows in Illinois and Indiana using our work and the PIOT generated for Illinois in previous work. We will also demonstrate the collaborative aspect of this computational tool by live demo of the tool during this talk, thus highlighting the scaling impact of collaboration through cloud-based platform in making sustainable manufacturing decisions.

References

[1] Shweta Singh; Lan Zhao; Jaewoo Shin; Venkata Sai Gargeya Vunnava (2021), "PIOT hub for Automated Generation of Regional PIOTs," *https://mygeohub.org/resources/piotpy.*

625295 Solid Oxide Fuel Cell – Thermal Energy Storage (SOFC-TES) Integrated System Modeling and Energy Efficiency Optimization.

Abstract:

Concerns over greenhouse gas emissions and their associated environmental challenges continue to increase. As a result, cleaner energy generation technologies, such as fuel cells, are receiving attention as the next generation's energy sources. One of these attractive energy technologies is the solid oxide fuel cells (SOFCs) due to its several advantages such as 1) reduced greenhouse gas emissions, 2) higher efficiency, and 3) flexible fuel selection compared to other types of fuel cells. However, the temperature of flue gas from SOFCs is significantly high due to its operating temperature between 600°C and 1000°C. Therefore, it is necessary to recover this waste heat so that we can maximize the process energy efficiency.

Many studies have been conducted to utilize the high-temperature exhaust gases from SOFCs so far. However, they only focused on the direct utilization of the heat through the electricity generation, using processes such as gas turbine[1], Rankine cycle[2], and combined heat and power (CHP)[3] in order to maximize the process energy efficiency. However, the aforementioned utilization methods result in losses where the generated electricity is further converted to other forms of energy such as thermal energy for space heating. The energy losses during this process are often significant.

Motivated by this, we propose the SOFC-integrated thermal energy storage (TES) method as a more efficient utilization technique for recovering waste heat from SOFCs. We employed a 4 kW-SOFC and combined it with the thermal energy storage system for storing the excess energy from the high-temperature exhaust gas. In order to integrate both processes, SOFC and TES mathematical models were developed using MATLAB. Next, the developed models are integrated, simulated, and the overall efficiency of the SOFC-TES model is compared to the general utilization methods. To further improve the proposed model's thermal efficiency, we developed a heat exchanger network using Aspen Energy Analyzer. Through the developed heat exchanger network, the integrated model was optimized by pinch analysis.

The optimization results revealed that the process efficiency of the integrated system, that is, SOFC-TES, showed superiority compared with the traditional methods of utilizing the waste heat from SOFCs. By this, excess heat from the exhaust gases can be reliably stored during low-demand seasons and reutilized during high-demand seasons to supplement residential and industrial energy demand or deficits. We concluded that the SOFC-TES could be widely used for various purposes due to its inherent characteristics, i.e., storing heat, such as a greenhouse in the smart-farm and human life, compared with the traditional method of waste heat utilization from the SOFCs.

References

[1] S.H. Chan, H.K. Ho, Y. Tian, **Modeling of simple hybrid solid oxide fuel cell and gas turbine power plant**, Journal of Power Sources, 109 (1) (2002), pp. 111-120

[2] Masoud Rokni, **Thermodynamic analysis of an integrated solid oxide fuel cell cycle with a Rankine cycle**, Energy Conversion and Management, 51 (2010), pp. 2724-2732

[3] S.H. Chan, C.F. Low, O.L. Ding, **Energy and exergy analysis of simple solid-oxide fuel-cell power systems**, Journal of Power Sources, 103 (2002), pp. 188-200

630767 Upcycling Food Waste for Hydrogen Storage: Technoeconomic Assessment

Abstract:

While annually 1.6 billion tons of food waste is estimated to be discarded globally by 2030, according to U.S. Environmental Protection Agency (EPA) the corresponding figure is projected forecast to reach approximately 66 million tons for U.S. With the monumental challenge of food waste management, researchers are seeking avenues of upcycling FW into higher value materials such as hydrogen storage medium. Therefore, the objective of this research was to evaluate the technoeconomic prospect of activated carbon production from food waste for the ultimate purpose of hydrogen storage through the application of hydrothermal carbonization of food waste followed by chemical activation of the hydrochar using KOH. The process was simulated using Aspen Plus and the process flow diagram (PFD) was constructed such that it mainly constituted of 2 rotary kilns and filter press, 1 rotary cooler, dryer, homogenizer, as well as other necessary equipment - pumps, heat exchangers, storage tanks. The facility processes 96 ton of food waste per day to be converted to activated carbon powder with a carbon yield of 20.4%. Also, it was assumed that the activated carbon manufacturing facility obtains food waste of Brevard County, Florida and the facility operates continuously 330 days a year. Several parameters were incorporated in the study including equipment sizing, capital costs and operating costs, such as labor, utilities, maintenance and equipment depreciation. Break-even cost from the economic analysis was evaluated and sensitivity analysis was carried out to reveal the effect of 7 different process variables on the break-even cost where raw materials' (KOH and food waste) and wastewater treatment cost significantly affected it. The unprecedented analysis shed light on the economic feasibility of producing activated carbon from food waste as a cost-effective route of upcycling it.

625694 Economic and Environmental Impact of Biochar Addition on Anaerobic Digestion Process

Abstract:

Anaerobic digestion (AD) technology is becoming a promising pathway for producing clean and renewable natural gas from waste resources. Despite growing industrial adoption, AD technologies still suffer from inefficiency due to process instability and microbial inhibitions. Scientists have demonstrated that biochar addition has positive impacts on AD. Biochar addition increases biogas yield and methane concentration and reduces gas contaminants. This influence is attributed to several mechanisms: providing support for microbial colonization, adsorbing excess inhibitory products, and improving electron transfer between syntrophic microorganisms. However, due to the cost and energy-intensive conversion process, the additional costs for biochar addition could offset the resulting energy benefits. This study investigates an integrated biorefinery where corn stover biochar is added to the AD of animal manure to produce biogas-derived Renewable Natural Gas (RNG) for pipeline injection.

A kinetic model of AD, biogas upgrading, and pipeline injection is simulated in Aspen Plus ™ for a mid-size farm with 5000 cattle heads. The process simulation model estimates a 28% increase in biogas yield for 3% (wt. % of wet mass) of biochar addition. High pressurized water scrubbing process is used for biogas upgrading into RNG. Techno-economic analysis (TEA) determines the minimum RNG selling price for a 30-year plant lifetime with a 10% Internal Rate of Return (IRR). TEA estimates the minimum RNG selling price as $42/MMBTU and $35/MMBTU for the without biochar and with biochar scenario, respectively. Biochar addition lowers the RNG selling price by 16% whereas adding revenue from digestate selling and renewable identification number (RIN) credits result in up to 32% reduction. Biochar addition increases the capital and operating costs by 8% and 13%, respectively. Cradle-to-gate life cycle analysis (LCA) is performed in SimaPro ™ to evaluate the GHG emissions of biochar amended AD for RNG production. GHG abatement for the without biochar and with biochar scenario is 0.038 kg $CO_{2\,eq}$/MJ of RNG and 0.123 $CO_{2\,eq}$ /MJ of RNG, respectively. LCA results show biochar addition leads to 2.18 times GHGs abatement compared to the without biochar scenario. Environmental benefits mostly come from the additional carbon sequestration with biochar land use and fossil fuel displacement by syngas from pyrolysis. Sensitivity analysis shows the influence of different major processes and economic parameters on the RNG selling price. The process is cost-prohibitive for small-scale farms due to the significant economy of scale penalty for the biogas upgrading and RNG injection facility. Increased biogas yield reduces the manure input leading to less digestate production. This would decrease the environmental benefits of emissions avoidance from conventional manure management and carbon sequestration per MJ of RNG while reducing the emissions from digestate transportation and applications. Uncertainty analysis considers the variations of key technical and economic parameters within a reasonable range and gives an estimated range of fuel selling prices and GHG abatement. Minimum RNG selling prices vary between $8.82/MMBTU and $16.28/MMBTU and GHG abatement vary between 0.06 kg $CO_{2\,eq}$/MJ and 0.16 kg $CO_{2\,eq}$/MJ within 10-90% confidence intervals.

627985 Optimizing Syngas Production Process By Downdraft Gasifier Using Response Surface Methodology

Abstract:

The gasification process of biomass to produce combustible gases has an attracted interest recently. The downdraft gasifier has an advantage over other types of gasifiers because of its small size, high quality gas production, and low tar yield. One of the challenges in the gasification process is the production of low-quality syngas (low concentration of H2 and CO) and large amount of tar. To overcome this problem, optimization of the gasifier operating conditions was employed to achieve high-quality product gases and hence high gasification process efficiency. In this work, the influence of initial gas temperature, airflow rate, and mole fraction of oxygen in primary air, on the gasification efficiency and product quality was optimized by response surface methodology. The obtained results showed that the reduction in the temperature caused a low concentration of carbon monoxide and a high amount of tar in the product, thereby there was a decline in the efficiency of the gasification process.

623925 Techno-Economic Evaluation of Nano-Porous Silica Production from Rice Husk and Sand

Abstract:

Global rice production in 2019 was estimated at about 500 million tons. Rice husk is a by-product from rice, including about 20 wt% of rice and is needed to be disposed as waste. Rice husk contains approximately 14-25% of ash. The silica content in rice husk ash (RHA) is above 90%. Nanoporous silica (NPS) used in industrial fields such as reinforced rubber additives (tires), concrete mixture (cement binder), nano-fibers, nano-filters, and membranescan be produced from RHA. Rice hisk is combusted at a relatively low temperature (650-800°C) and RHA is produced, whereas the silica sand is operated at 1400-1500°C to produce silicate. In addition, compared to the conventional NPS production from sand, The NPS production from RHA has an advantage in lowering the production cost due to low material cost and energy consumption.

In this study, the techno-economic feasibility for the sodium silicate ($Na_2O{\cdot}xSiO_2$)-based nano-porous silica (SS-NPS) production from sand and rice husk was performed. The return on investment (ROI) and payback period (PBP) were evaluated according to the plant size for the two cases. The effects of raw material and product prices on ROI and PBP were analyzed and economic conditions to be economically viable were suggested.

AIChE
The Global Home of Chemical Engineers

A CAPITAL AND ENERGY EFFICIENT ALTERNATIVE EMISSIONS CONTROL TECHNOLOGY FOR CONTROLLING EMISSIONS USING BEAD ACTIVATED CARBON

Kim K. Tutin, Captis Aire LLC, East Point, GA United States
Harold R. Cowles, Environmental C&C Inc., Scottsdale, AZ United States
John M. Berger P.E., Environmental C&C Inc., Fishers, IN United States

TABLE OF CONTENTS:

I. INTRODUCTION
II. DESCRIPTION OF FBC TECHNOLOGY: FOUR PRIMARY PARTS
III. ADSORBENT MEDIA
IV. OPTIONS
- RECOVERY SYSTEM
- OXIDATION SYSTEM

V. PROCESS PERFORMANCE CHARACTERISTICS
VI. AWARDS
VII. APPLICATIONS
- VALUE CREATION CALCULATION EXAMPLE

VIII. CASE STUDY: FBC TECHNOLOGY SPECIFIC TO WOOD DRYING
- DEMONSTRATION DESCRIPTION
- RESULTS OF AIR POLLUTION CONTROL COMPLIANCE TESTING
- RESULTS OF VOC REDUCTION EFFICIENCY PERFORMANCE MEASURED BY FLAME IONIZATION DETECTOR (FID):
- RESULTS OF VOLATILE ORGANIC COMPOUNDS CAPTURE
- VALUE OF TERPENES
- TERPENES TYPICAL PROPERTIES
- MAP OF POTENTIAL WOOD DRYING SITES FOR FBC USE
- RESULTS OF DISCUSSIONS WITH ENVIRONMENTAL AGENCIES
- FINAL RESULTS FROM WOOD DRYING DEMONSTRATION

IX. CONCLUSIONS
X. ACKNOWLEDGEMENTS

I. INTRODUCTION:

In manufacturing, process exhaust gas (air) streams released may contain volatile organic compounds (VOCs), Hazardous Air Pollutants (HAPs), particulate, etc. These are considered pollutants/emissions and must be removed from the air prior to release to avoid air pollution emissions into our communities. Add-on controls either collect the VOCs for recovery and reuse or destroy the VOCs. A number of control options are available including thermal oxidizers, catalytic oxidizers, and biofiltration which destroy pollutants. Other control systems including condensation, scrubbing, and adsorption systems may recover pollutants.

This paper provides an introduction to an adsorption technology, the Fluidized Bed Concentrator (FBC) with Bead Activated Carbon (BAC) media. It describes 1) the FBC technology, 2) the BAC media used in these systems, 3) the performance characteristics, 3) typical applications of this technology, and 4) a case study review of the FBC technology demonstration for controlling emissions and collecting terpenes from wood drying.

II. DESCRIPTION OF FBC TECHNOLOGY: FOUR PRIMARY PARTS

1. Adsorber:

The adsorber section has trays filled with very adsorbent BAC media. Contaminated process exhaust enters from the bottom, passes upward through the adsorption trays, and fluidizes the Bead Activated Carbon (BAC) adsorbent media. When the process air interacts with the BAC, it adsorbs (captures) the terpenes Volatile Organic Compounds (VOCs) in its pores. VOC saturated BAC flows to the bottom of the adsorber vessel and is transferred using fans to pneumatically transfer it to the desorber. Cleaned (reusable) BAC from the desorber feeds continuously into the top of the adsorber, providing counter-current VOC removal. Cleaned air exits out the top of the adsorber.

2. Desorber:

In the desorber, the BAC media is heated (regenerated), causing it to release the VOC vapors into a low volume, gas stream. This gas stream is then circulated continuously to the next stage in the process. The regenerated (cleaned) BAC is then returned continuously to top of the adsorber for reuse. Returning this regenerated BAC to the top of the adsorber provides a final "polishing" of emissions to remove residual emissions prior to release of the process exhaust air.

Desorber Recovery System: In the recovery system, a low volume inert carrier (nitrogen) gas is introduced into the bottom of the desorber. Gas flows up through the desorber as the BAC flows downward. The system is heated via either hot oil or electric heating. The heat causes the VOCs to volatilize off the BAC. The VOC vapors are liberated resulting in a concentrated waste flue gas stream that moves from the desorber to the condenser. This system may eliminate the on-site use of natural gas entirely. Thus, it offers an "All Electric" control technology.

"All Electric" Option

Desorber Oxidation System: In the oxidation system, clean hot stripping air at 100-700°F is introduced to the bottom of the desorber. It flows upward across the BAC media to both volatilize the VOCs and fluidize the BAC. The VOC vapors are liberated resulting in a concentrated waste flue gas stream that moves from the desorber to the oxidizer.

3. Condenser or Oxidizer:

The gas stream from the desorber is circulated continuously to this system for further treatment to either capture and recover VOCs or oxidize them.

"All Electric" Option

Condenser Recovery System: In the recovery system, nitrogen gas circulates from the desorber to the condenser continuously. The VOC vapors are sufficiently concentrated such that the vapors can be cooled via a tube and shell condenser converting the vapor to liquid. The liquid drains continuously into a storage system. The liquid VOCs can then be reused or sold. The nitrogen recirculates back to the desorber where it entrains the VOCs vapors again and the cycle continues.

Oxidizer Oxidation System: In the oxidation system, the vapors are instead sent to a small thermal oxidizer to treat the VOCs. While this system does not provide recovered (collected) VOCs, it nevertheless retains FBC technology benefits including significantly reducing natural gas usage. The burner component of the thermal oxidizer is selected to provide enough energy to increase the thermal oxidizer to the 1400°F operating temperature. The burner is designed to minimize the "fire rate and thus consume negligible fuel. Optionally, a "low NO_x" design can be selected to minimize production of NO and NO_2 (oxides of nitrogen). Thus, the system minimizes energy usage and GHG emissions. [1]

4. Side Stream Reactivator (SSR) [2]:

A small portion of the BAC media exiting the desorber may be transferred to an optional SSR. The SSR reactivates the BAC by heating it to 1450°F with the introduction of a small amount of water which reactivates the BAC continuously to regenerate it and thus maintain its ability to adsorb VOCs. Reactivated BAC from the SSR is transferred back to the top of the adsorber for reuse.

III. ADSORBENT MEDIA:

The BAC adsorbent media utilized in this system is an essential element of the FBC technology. Specialized Bead Activated Carbon (BAC) efficiently adsorbs VOCs (i.e. terpenes) from the process air. The BAC has high-fill (adsorption) capacity, is a small sphere-shaped particle that is less than 1 mm in diameter with a narrow particle size distribution, has low dust, has high purity, has high flowability, and has high strength and wear resistance as measured by ASTM D 3802 – 79 (2005) Standard Test Method for Ball-Pan Hardness of Activated Carbon.[3] The spherical shape enables the BAC to fluidize and flow.

The BAC media can be reactivated and reused. This is accomplished either by either continuous or batch processes.

---

[1] "VOC Removal and Destruction System and Method". U.S. Patent and Trademark Office. (USPTO) https://appft.uspto.gov/netacgi/nph-Parser?Sect1=PTO2&Sect2=HITOFF&p=1&u=%2Fnetahtml%2FPTO%2Fsearch-bool.html&r=1&f=G&l=50&co1=AND&d=PG01&s1=20190022579.PGNR.&OS=DN/20190022579&RS=DN/20190022579 (Accessed October 4, 2021).

[2] "GASTAK® Solvent Recovery and Odor Removal by Fluidized Bed Process". Kureha Ecology Management Co. , LTD website. https://www.kurekan.co.jp/en/download/gastak.pdf (Accessed September 18, 2021).

[3] "ASTM D3802-79 (2005) Standard Test Method For Ball-Pan Hardness Of Activated Carbon. https://webstore.ansi.org/standards/astm/astmd3802792005 (Accessed September 18, 2021).

Continuous Reactivation: In the continuous process, a small amount of BAC media is diverted from the BAC exiting the desorber and sent to an in-line side stream reactivator (SSR).[4] It is then returned to the top of adsorber. Returning this reactivated BAC to the top of the adsorber provides a final "polishing" of emissions to remove residual emissions prior to release of the process air.

Batch Reactivation: In the batch process, for operators that do not have an SSR, the BAC media is unloaded periodically, shipped offsite for reactivation, and then reloaded once returned. In the interim, a second load of BAC is utilized. Reactivation is typically performed by California Carbon who has been providing reactivation services since 1975.[5] This service includes a reactivation warranty *"Carbon will be reactivated to 95% of Virgin Activity per ASTM D 3467 and Carbon will lose no more than 10% of incoming volume of the specified mesh size in the reactivation process."*[6] BAC lifetime is site and application specific. Some operators have been reusing their BAC for years.

Bead Activated Carbon (BAC) Adsorbent Media

Bead Activated Carbon (BAC), the Adsorbent Media, Fluidized in Tray

IV. FBC OPTIONS: RECOVERY AND OXIDATION

The Fluidized Bed Concentrator can be utilized to either recover the VOCs or oxidize them as shown in the figures. Operators typically select the desired option based on a number of factors especially including the anticipated value and reusability of the recovered VOCs. The value of the VOCs depends on a number of factors including the anticipated quantity recovered, purity, salability, and/or reusability.

Recovery: When the value of the recovered VOCS is high and if the cost of recovery is less than the cost of purchasing new material, then recovery makes sense.

[4] "GASTAK® Solvent Recovery and Odor Removal by Fluidized Bed Process". Kureha Ecology Management Co. , LTD website. https://www.kurekan.co.jp/en/download/gastak.pdf (Accessed September 18, 2021).
[5] "California Carbon Co. Inc. Welcome". California Carbon website. http://www.californiacarbon.com/index.html (Accessed September 18, 2021).
[6] "Custom Spent Carbon Reactivation". California Carbon website. http://www.californiacarbon.com/pdfs/Custom%20Reactivation.pdf (Accessed September 18, 2021).

Oxidation: If the VOC has low or no recovery value, as, for example, when it is a mixture, or if it poses disposal concerns, such as for toxic compounds, then destruction probably makes the most sense. Excluding value derived from VOC recovery, the oxidation option retains FBC system performance characteristics including reduction in energy usage.

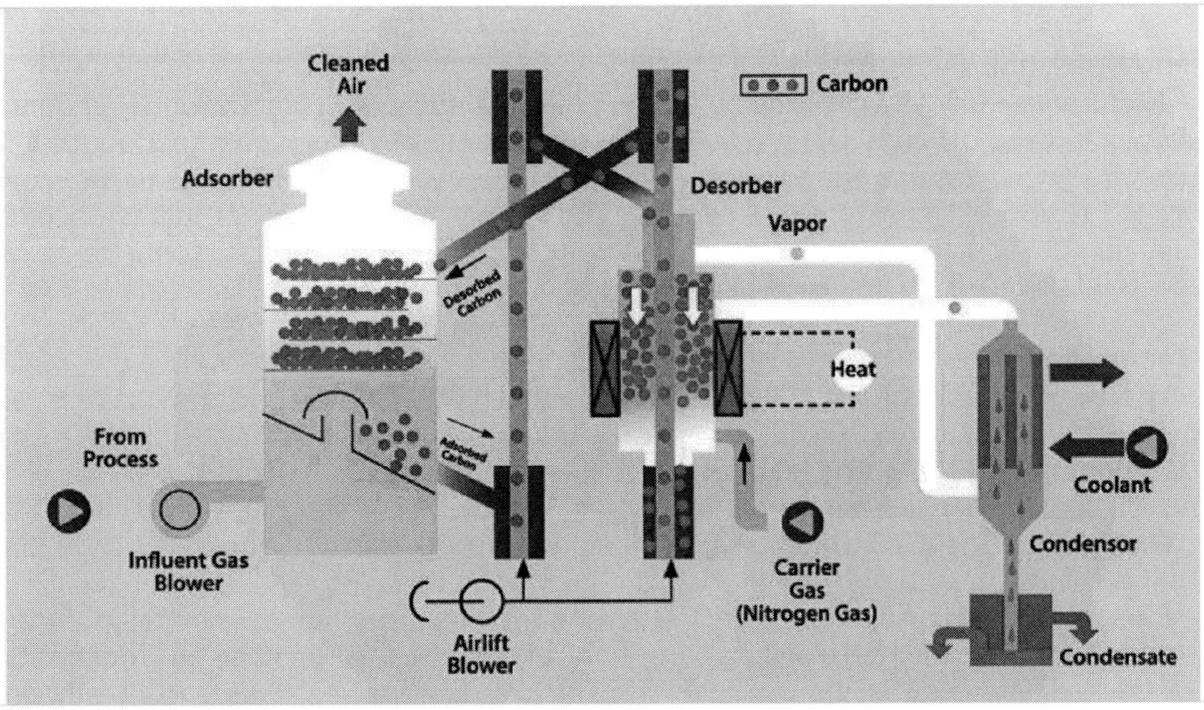

RECOVERY SYSTEM: Adsorbs, Desorbs, and Condenses VOCs

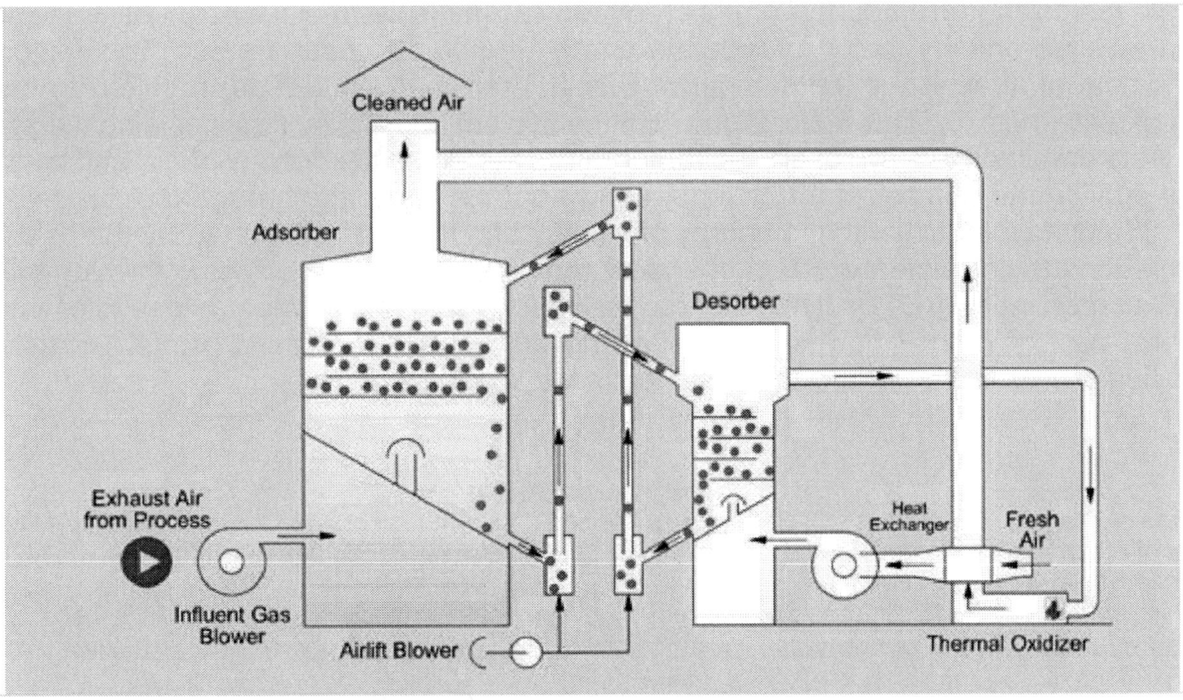

OXIDATION SYSTEM: Adsorbs, Desorbs, and Oxidizes VOCs

V. PROCESS PERFORMANCE CHARACTERISTICS:

1. <u>The FBC reduces energy usage (i.e. natural gas) versus oxidation systems due to reduction or elimination of energy used to heat the air going through thermal oxidizers:</u>

Baseline Oxidation Systems:

Thermal oxidizers burn VOCs by heating them in enclosed combustion chambers in the presence of excess oxygen. The precise temperature and residence time needed for effective reduction of VOC emissions varies with the nature of the VOC. However, temperatures of 1400-2000°F and residence times of 0.5-2 seconds are typical. Since auxiliary fuel is necessary to maintain combustion temperatures, most thermal oxidizers incorporate some form of heat recovery to lower auxiliary fuel costs.

FBC Technology:

The air going through the adsorber is not heated. Thus, the natural gas usage is reduced or eliminated.

Recovery – "All Electric" Option: For a recovery system, the control device can be "All Electric". It could eliminate the use of natural gas entirely.

Oxidation: The FBC oxidation system significantly reduces natural gas usage but does not eliminate it.

Energy Reduction Calculation: The amount of energy that is eliminated, that would have been used by the thermal oxidizer, can be calculated utilizing one of many credible sources. One of the best sources is the Institute of Clean Air Companies' Guidance Methodology[7]. Calculators are also available on-line including those provided by the EPA[8], Cycle Therm[9], and Captis Aire[10]. With these calculators, users can input their specific data and the calculators will calculate the estimated amount of energy required to run the thermal oxidizer. All these sources rely on a single primary factor to calculate energy required to heat the air. This factor is the specific heat of air which is the number of British Thermal Units (BTUs) required to heat one pound of air by one °F. This factor along with the system specific factors including the standard cubic feet per minute (sCFM) of air, the temperature increase, the density of air, and conversion factors provide the number of BTUs required to heat the air going through the device. These factors are used together with the energy efficiency and the heat generated by the oxidation of the emissions to calculate the BTUs of energy required to heat the air going through the RTO. Users can then multiply BTUs times the forecasted long-term cost of natural gas to estimate the long-term cost of natural gas required over the 20+ year life of the thermal oxidizer. This calculation provides the value of the natural gas elimination.

---

[7] "Institute of Clean Air Companies (ICAC) Guidance Method for Estimation of Gas Consumption in a RTO". Institute of Clean Air Companies website. https://cdn.ymaws.com/www.icac.com/resource/resmgr/RTO-F1.pdf (Accessed September 18, 2021)

[8] "Air Pollution Control Cost Estimation Spreadsheet". Environmental Protection Agency (EPA) website. https://www.epa.gov/sites/production/files/2018-01/oxidizers_calc_sheet_finalversion_1-16-2018.xlsm (Accessed Feb 3, 2021).

[9] "Operating Cost Calculator". Cycle Therm website. http://www.cycletherm.com/resources/operating-cost-calculator.aspx (Accessed Feb 3, 2021).

[10] "Calculate Engineering Estimate of Natural Gas Usage and Greenhouse Gas Generation". Captis Aire website. https://www.captisaire.com/Value-Calculator/ (Accessed Feb 3, 2021).

2. The FBC generates revenue from sales of recovered organics:

When using a recovery system, including the FBC, operators have the option to collect and sell or reuse the volatile organic compounds recovered. The liquid VOCs are collected from the FBC condenser for reuse or sale. The value of recovered VOCs will depend on the type, amount, and purity of VOCs collected.

3. The FBC reduces Greenhouse Gases (GHG) emissions versus oxidation systems due to the reduction or elimination of burning natural gas:

Due to the reduction or elimination of natural gas usage and reduction in electrical usage by the FBC, greenhouse gas emissions are also reduced. The greenhouse gas emissions reduction can be calculated based on the amount of carbon in natural gas eliminated and its stoichiometric conversion to carbon dioxide (CO_2). The Environmental Protection Agency (EPA) has an on-line calculator[11] that enables the user to calculate greenhouse gas emissions that are eliminated by eliminating natural gas used to heat the air in the thermal oxidizer.

4. The FBC enables creation of carbon offsets versus oxidation systems:

In conversations with American Carbon Registry[12] and ClimeCo[13], it may be possible to measure, register, and verify carbon offsets for a site converting to the FBC technology from an oxidation system due to the reduction in energy usage.[14]

The quantification methodology and value of carbon offsets is evolving quickly. But these carbon offsets can be sold or traded externally. Or they may be utilized for internal credit accounting purposes. Potential values for these carbon offsets vary widely.[15] However, as of early 2021, American Carbon Registry and ClimeCo estimated the value of carbon offsets for an FBC project at $5/ton.[16]

5. The FBC reduces electricity usage versus oxidation systems:

The horsepower required to run the fans to move the process exhaust air through the BAC media is reduced for the FBC due to the fact that it's BAC is fluidized (moving) rather than packed (non-moving). The pressure drop across the system is reduced from a typical 10-20" to 3.5". Thus, the horsepower required to run the fans for the FBC may be ~ ¼ of that required for the RTO.

To estimate the electrical savings, the sum of electrical demands from the FBC is subtracted from the sum of electrical demands of the thermal oxidizer. Net kW·h along with site-specific electricity costs and operating hours per year are used in combination to estimate electrical savings.

---

[11] "Greenhouse Gases Equivalencies Calculator - Calculations and References". EPA website. https://www.epa.gov/energy/greenhouse-gases-equivalencies-calculator-calculations-and-references (Accessed December 28, 2020).

[12] "Harnessing the Power of the Markets to Improve the Environment". American Carbon Registry website. https://americancarbonregistry.org/ (Accessed February 3, 2021).

[13] "Scalable greenhouse gas projects to earn real financial returns for your business." ClimeCo website. https://climeco.com/ (Accessed February 3, 2021).

[14] "AM0044: Energy efficiency improvement projects - boiler rehabilitation or replacement in industrial and district heating sectors --- Version 2.0.0". Clean Development Mechanism (CDM) website. https://cdm.unfccc.int/methodologies/DB/3HZ4USHZ2W449HMAXZN420E5PJB1QF (Accessed Sept 18, 2021)

[15] "Allowance Price Explorer". International Carbon Action Partnership website. https://icapcarbonaction.com/en/ets-prices (Accessed Jan 20, 2021).

[16] Eric Ripley (Director Industrial Programs, American Carbon Registry), in a phone conversation with Kim Tutin November 17, 2020.

6. **The FBC enables superior mass transfer of VOCs through counter current flow fluidized BAC media that is continuously regenerated:**

Baseline Adsorption Systems:

Adsorption: In sorbent treatment systems, process exhaust air moves across the sorbent. VOCs are removed via attraction (adsorption) of the VOCs on carbon or zeolite sorbent media in packed (non-moving) beds. This may be accomplished by concentrators that have wheels with sectors containing activated carbon or zeolite media. In this arrangement, most of the space on the wheel functions in adsorption mode, with one small sector being regenerated at any given time. Or the process exhaust air may be forced through large vessels containing sorbent media. In this arrangement, the VOCs are treated via a once-through process.

Regeneration / Reactivation: Regeneration is accomplished by heating the sorbent media with heaters, heated gases, steam, etc. If using steam, water may be a byproduct. This can be accomplished either via off-site regeneration or by alternating the flow between multiple vessels where one vessel functions in adsorption mode while the other is being regenerated at any given time. In both these arrangements, the sorbent media is being regenerated in batch or semi-continuous fashion. Reactivation may be accomplished by sending the sorbent media off site for reactivation. Or the sorbent media may be discarded when it becomes saturated.

Effect of water vapor: Especially for process exhaust air heavily laden with water vapor, i.e. 10% B_{ws} moisture by volume[17] or more, the sorbent media may become partially or fully saturated with water. At high moisture content, water molecules begin to compete with the hydrocarbon molecules for active adsorption sites. This reduces the capacity and the efficiency of the adsorption system.

FBC Technology:

Adsorption: The BAC media in the fluidized bed concentrator adsorber is fluidized (moving) by the process air moving opposite (counter current) to the BAC as it moves downward through the trays in the adsorber. This provides excellent mixing of the BAC and contact with the air. As this occurs, VOCs or odors are transferred from the process air to the BAC. The cleaned air exits through the top of the adsorber.

Regeneration / Reactivation: The BAC media is regenerated continuously via continuous recirculation to and from the desorber. It is reactivated continuously if system includes the Side Stream Reactivator (SSR). This ensures the BAC retains the capacity to maintain adsorption efficiency. This is done at temperatures well above the boiling point of water, often around 400°F or more. This minimizes or eliminates accumulation of water and/or organics on the BAC.

Effect of water vapor: Continuous regeneration and reactivation at temperatures above the boiling point of water, i.e. 400°F, volatilizes the water off the BAC media and prevents the BAC from becoming saturated with water. Thus, it maintains adsorption efficiency even when treating process exhaust streams with high moisture contents, as are typical for wood drying applications, i.e. ≥20% B_{ws} moisture by volume[18]. This system minimizes or eliminates need for dehumidification systems.

---

[17] "METHOD 4—DETERMINATION OF MOISTURE CONTENT IN STACK GASES". EPA website. https://www.epa.gov/sites/default/files/2016-06/documents/m-04.pdf. (Accessed October 17, 2021).
[18] "METHOD 4—DETERMINATION OF MOISTURE CONTENT IN STACK GASES". EPA website. https://www.epa.gov/sites/default/files/2016-06/documents/m-04.pdf. (Accessed October 17, 2021).

Effect of Fluidized (moving) vs Packed (non-moving) media: BAC media fluidization prevents pockets of high concentration VOCs, provides higher dry particulate tolerance (reduced plugging of interstitial spaces between media), and enables continuous on-site regeneration and reactivation.

TABLE I
PHYSICAL AND OPERATIONAL COMPARISON OF THE FBC VS THE RTO

Parameters:	RTO	FBC	Advantages of FBC
Treatment Temperature, °F	>1400°F	175°F	• Adsorption at process air temperature • No natural gas required for heating
Media Desorption Temperature, °F	>1400°F	400°F	• Much less energy required for desorption • Continuous media regeneration, above the boiling point of water, prevents the media from getting saturated with water. Thus, it maintains its ability to adsorb pollutants.
Media depth, feet (ft)	16 ft	1 ft	• Less Electricity (fan HP) required to move air through less media
Media Configuration	Packed (Not Moving)	Fluidized (Moving)	• Less Electricity (fan HP) required to move process air through fluidized media
Pressure Drop, Inches (")	10-20"	3.5"	• Less Electricity required to move process air against less pressure
Fan size, Horsepower, (HP)	2400 HP	600 HP	• Less HP due to media that is fluidized. • Less susceptible to plugging from particles
Emissions Treatment	Oxidation to CO^2	Adsorption (Capture)	• Emissions are NOT oxidized to CO_2, so reduces GHG emissions • Emissions collected can be sold or reused
Media Reuse	Discarded	Reused	• Media desorption with 'in unit' continuous regeneration. • Flowable bead shaped media can be removed and reused
Safety		Inherently Safer	• Lower operating temperatures • No large heavy moving parts • Easier cleaning and maintenance

VI. AWARDS:

The FBC has been recognized for excellence in environmental performance. Honda Motors of Alabama was recognized as *"Air Conservationist of the Year"* when they implemented the FBC to control emissions when they increased production.[19] Ford's "Fumes to Fuels" work was awarded the EPA's *"Clean Air Excellence"* award.[20] In this work, they utilized the FBC to capture solvent-laden air, strip the solvent from the air, concentrate the solvent up to 2,000 times, and send it to a fuel reformer. The fuel reformer used the concentrate to generate hydrogen-rich fuel, which was directed into a fuel cell to generate electricity.

VII. APPLICATIONS:

Concentration systems, including the FBC, are advantageously used in front of control devices when treating very high volumes of very dilute exhaust gas to reduce energy usage. Recovery systems are beneficial when the value of the recovered VOCs is high. The following are a few examples where the FBC technology has been utilized historically.

TABLE II
FBC APPLICATIONS

Application	Chemistry
Adhesive Tape Mfg	Toluene
Agricultural Products	Organic Acid & Ammonia Odor
Aluminum Casting	Phenol, Formaldehyde, Ammonia
Animal Lab	Odor Control
Atomic Power Plant	Styrene
Auto Parts Degreasing	Trichloroethylene
Bakery	Odor Control
Brewery	Odor Control
Car Wash	Kerosene
Cellophane Coating	Toluene, Ethyl Acetate, Butyl Acetate
Ceramic Condenser	Terpineol
Confectionery	Odor control
Chemical Production	Toluene, Higher Molecular Weight Alcohols, Carbon Tetrachloride

[19] "HMA Honored as Air Conservationist of the Year". Honda website. https://www.hondaalabama.com/article/hma-honored-as-air-conservationist-of-the-year (Accessed September 21, 2021).
[20] "Clean Air Excellence Award Recipients". Environmental Protection Agency (EPA) website. https://www.epa.gov/sites/default/files/2015-06/documents/clean_air_excellence_award_recipients_year_2003.pdf (Accessed September 21, 2021).

TABLE II (CONTINUED)
FBC APPLICATIONS

Application	Chemistry
Dye Production	Perchloroethylene
Electronics Mfg Plant	DMF, MEK, etc.
Electronics Parts Cleaning	Trichloroethylene, Toluene, 1,1,1-Trichloroethane
Fabric Washing	Perchloroethylene
Feed Processing	Odor Control
Film Coating	MEK, Methyl Cellosolve, Toluene, Tetrahydrofuran
Film Laminating	Toluene, Ethyl Acetate, n-Hexane
Gravure Printing	Toluene, Xylene, n-Propanol, n-Propyl Acetate, mixed
Iron Casting	Phenol, Formaldehyde, Ammonia
Lacquering of Film	Toluene, Ethyl Acetate, MEK
Landfill	Odor Control
LSI Mfg Plant	Phenol, Acetone, Methanol, Mixed Solvents
Lubrication Mfg	Mixed Solvent Odor Control Magnetic Tape, Chlorinated Hydrocarbon
Medicine Mfg Plant	Butanol, Methylene Chloride
Paint Booth	Thinner
Printing Ink Mfg	Mixed Solvents
PVC Resin Mfg Plant	Vinyl Chloride
Resin Mfg	N-Hexane, Methylene Chloride
Rubber Coating	Toluene
Rubber Vulcanizing	Odor Control
Sandpaper Mfg	Toluene, Xylene, Ethyl Acetate
Semiconductor Mfg	IPA, HMDS, NMP, Ethanol, Methanol, Ethyl lactate
Silk Screen Printing	Xylene, Mixed Solvents
Teflon Tubing Mfg	Perchloroethylene
Wastewater Aeration	Odor Control

TABLE III
EXAMPLES OF FBC SYSTEMS INSTALLED

FLUIDIZED BED CONCENTRATOR SYSTEMS INSTALLED: [21]	
Industry	Semiconductor Chip Manufacturing
Location	Israel
Inlet Stream	20,000 CFM
VOCs	Semiconductor solvents, 50 ppmv
Performance Achieved	95% Reduction Efficiency
Notes	System located in populated area of a major city has eliminated nuisance odor
Industry	Semiconductor Chip Manufacturing
Location	Arizona
Inlet Stream	36,000 CFM
VOCs	Semiconductor solvents, 25-100 ppmv
Performance Achieved	90%+ Reduction Efficiency
Notes	System met strict local compliance requirements while achieving Zero Unscheduled Downtime (ZUD)
Industry	Teflon Tubing Manufacturing
Location	Massachusetts
Inlet Stream	1,800 CFM
VOCs	Perchloroethylene, 1500 ppmv
Performance Achieved	98% Reduction Efficiency
Notes	Recovered solvent re-used in process. System payback in 1 year.
Industry	Furniture Painting
Location	California
Inlet Stream	50,000 CFM
VOCs	Paint Solvents, 25-50 ppmv
Performance Achieved	90%+ Reduction Efficiency
Notes	Recovered solvent and reformulated with paint
Industry	Military Vehicles
Location	Michigan
Inlet Stream	150,000 CFM
VOCs	Paint Solvents
Performance Achieved	95% Reduction Efficiency
Notes	Controlled multiple different interchangeable VOC sources.

[21] "Environmental C&C Inc". Environmental C&C website. https://www.environmentalcc.com/ (Accessed September 18, 2021).

Automotive Site, Alabama, 100,000 CFM
Paint Solvents, >90% Reduction Efficiency
Two adsorbers and one desorber

Image shows three adsorbers one desorber	
Image shows ductwork that enabled control of any of the 3 sources to any of the 3 adsorbers. **Each adsorber has its own process fan.** **System provided control of multiple different interchangeable VOC sources**	

Military Vehicles, Michigan, 150,000 CFM
Paint Solvents, 95% Reduction Efficiency

TABLE IV: VALUE CREATION CALCULATION EXAMPLE

Fluidized Bed Concentrator (FBC) vs Regenerative Thermal Oxidizer (RTO)	Value per Year
1) Reduce Energy Usage (Natural Gas) due to Elimination of Energy Used to Heat the Air in the RTO: [22] [23] a. $\frac{mmBTU}{Year}\ to\ heat\ air\ in\ RTO = (\Delta\,T)\ (sCFM)(1.08)(\%Heat\ Loss)(\frac{Operating\ Hours}{Year})/1000000$ b. $(1500°F - 175°F) * (300{,}000sCFM) * \left(\frac{1.08\ BTUs}{sCFM \cdot Hour \cdot Degree°F}\right) * (7.5\%) * (8000\frac{Hrs}{Yr})/1000000$ c. $= \left(257{,}580\frac{mmBTU}{Yr}\right) - [\left(250\frac{lbs\ VOC}{Hour}\right) * \left(15{,}000\frac{BTUs}{lb\ VOCs}\right) * \left(\frac{8000\ Hrs}{Year}\right)] = \mathbf{227,580}\frac{mmBTUs}{Year}$ d. $\left(\frac{227{,}580\ mmBTUs}{Year}\right)(\$3.5\frac{Long\ Term\ Cost\ of\ Natural\ Gas}{mmBTU}) = \mathbf{\$796,530\ Savings/Year}$	Save by Eliminating Natural Gas **$796,530**
2) Reduce Greenhouse Gases (GHG) Generated from Burning Natural Gas: [24] a. $Tons\ GHG = (mmBTU) * \left(\frac{117\ lbs\ CO2}{mmBTU}\right) * (\frac{1\ Metric\ Ton}{2205\ lbs})$ b. $\left(227{,}580\frac{mmBTUs}{Yr}\right) * \left(117\ lbs\frac{CO2}{mmBTU}\right) * \left(1\frac{Metric\ Ton}{2205\ lbs}\right) = \mathbf{12,076}\frac{Tons\ GHG\ Reduction}{Year}$	Sell Carbon Offsets **$60,378**
3) Sell Carbon Offsets (Carbon Credits) Utilizing Carbon Accounting Service Providers, i.e. ClimeCo: [25] a. $\left(\frac{12{,}076\ Tons\ GHG}{Year}\right) * \left(\frac{\$5}{Ton\ GHG}\right) = \mathbf{\$60,378/Year\ Selling\ Carbon\ Offsets}$	
4) Reduce Electricity Usage by Reducing Fan Power to Move Process Air through Fluidized Media: $(2400RTO - 600FBC\ HP\ fans - 800\ HP) * \left(\frac{0.75\ kW \cdot h}{HP}\right)\left(\frac{\$0.07}{kW \cdot h}\right)\left(\frac{8000\ Hrs}{Year}\right) = \$420{,}000\frac{Savings}{Year}$	Save by Reducing Electrical Usage **$420,000**
5) Sell Terpenes [26] **Yielded from Green Pine Wood** [27]**:** a. $\left(\frac{1\ Gallon\ Terpenes}{Oven\ Dry\ Ton\ Wood}\right) * \left(\frac{300{,}0000\ Dry\ Tons\ Wood}{Year}\right) * (75\%Yield) * \left(\frac{\$3}{Gal}\right) = \mathbf{\$675,000/Yr}$ **Selling Terpenes**	Sell Terpenes **$675,000**
6) Reduce Maintenance Costs:[28] Media changeouts, large heavy moving part replacement, corrosion repairs, etc.	Save up to **~$1 Million**
Total Value Creation per Year per Site:	**Up to $3 Million**

---

[22] ICAC Guidance Method for Estimation of Gas Consumption in a RTO. Institute of Clean Air Companies website. https://cdn.ymaws.com/www.icac.com/resource/resmgr/RTO-F1.pdf (Accessed September 18, 2021)

[23] "Operating Cost Calculator". Cycle Therm website. http://www.cycletherm.com/resources/operating-cost-calculator.aspx (Accessed November 21, 2021).

[24] "Greenhouse Gases Equivalencies Calculator - Calculations and References". Environmental Protection Agency (EPA) website. https://www.epa.gov/energy/greenhouse-gases-equivalencies-calculator-calculations-and-references (Accessed December 28, 2020).

[25] "Allowance Price Explorer". International Carbon Action Partnership website. https://icapcarbonaction.com/en/ets-prices (Accessed Jan 20, 2021).

[26] "Price Chart for Gum Turpentine". PineChem website. http://www.pinechem.net/images/gall/ChartGNE.GIF Accessed Jan 3, 2021).

[27] "Sulfate Turpentine Recovery". Pulp Chemicals Association. Pages 9-12, 127. https://www.pinechemicals.org/store/viewproduct.aspx?id=1579188 (Accessed Jan 2, 2021).

[28] "Air Pollution Control Technology Fact Sheet". EPA website. https://www3.epa.gov/ttncatc1/dir1/fregen.pdf (Accessed October 17, 2021).

VIII. CASE STUDY: FBC TECHNOLOGY SPECIFIC TO WOOD DRYING

DEMONSTRATION DESCRIPTION:

Captis Aire LLC founder and Environmental C&C Inc. worked on a project together from 2015-2021 with a commercial oriented strand board (OSB) manufacturer to demonstrate the ability to utilize the FBC to control emissions from wood drying.

In wood products manufacturing, process gas exhaust streams released from drying wood contain VOCs, including turpentine (terpenes). These VOCs are considered pollutants/emissions and must be removed from the air, prior to release, to avoid air pollution emissions into our communities. Wood drying operations utilized in the manufacture of oriented strand board (OSB), lumber, wood to biofuels, i.e. wood pellets, plywood, particleboard, etc. are few examples of these processes that have process exhaust streams containing VOCs.

Work on this patent[29] pending[30] technology was executed in three phases. In the first phase, laboratory screening work was done to explore whether emissions from wood drying, primarily terpenes, could be adsorbed by Bead Activated Carbon (BAC) media at the expected operating temperatures. In the second phase, a small-scale pilot adsorber was installed at a commercial Oriented Strand Board (OSB) manufacturing site to demonstrate ability to adsorb emissions from wood drying under commercial manufacturing conditions. In the third phase, the pilot unit capabilities were expanded to include the adsorber, desorber, and condenser. The SSR performance was simulated by semi-continuous additions of virgin BAC during air emissions compliance testing.

In the third testing phase, the pilot unit ran for 6 months with over 1700 hours on-line. Ability to meet air emissions control compliance, per the Plywood and Composite Wood Products Maximum Achievable Control Technology (PCWP MACT) standard, was determined by two independent certified air emissions[31] stack testing companies[32] during two separate testing events.

---

[29] "METHODS AND SYSTEMS FOR RECOVERING TERPENE COMPOSITIONS FROM WOOD DRYING EXHAUST". World International Patent Office (WIPO) Website. https://patentscope.wipo.int/search/en/detail.jsf?docId=WO2019148208 (Accessed September 18, 2021).

[30] "METHODS AND SYSTEMS FOR CONTROLLING EMISSIONS FROM WOOD DRYING PROCESSES". WIPO Website. https://patentscope.wipo.int/search/en/detail.jsf?docId=WO2019148207 (Accessed September 18, 2021).

[31] "AIR Advanced Industrial Resources". AIR website. https://airtest1.com/ (Accessed April 16, 2020).

[32] "Testing and Modeling". John Zink Hamworthy Combustion website. https://www.johnzinkhamworthy.com/testing-modeling/ (Accessed Feb 3, 2021).

TABLE V
RESULTS OF AIR POLLUTION CONTROL COMPLIANCE TESTING (FBC UNIT):

Emissions Measured:	FBC %Reduction Efficiency[33]
Total Hydrocarbons (THC), as carbon including Methane	92%
Volatile Organic Compounds (VOCs), as carbon. (Total non-methane hydrocarbons – THC minus Methane)	92%
Formaldehyde	83%
Methanol	75%
GHC - Nitrogen Oxides (NOx)	55%
Particulate Matter (PM), total filterable	50%
Condensable Particulate Matter	59%
Total Particulate Matter, Filterable + Condensable	51%
Total PM2.5, Particle Size Analyses (PSA) Filterable + Condensable	60%
Total PM10, Particle Size Analyses (PSA) Filterable + Condensable	64%
Carbon Monoxide (CO)	7%
Sulfur Dioxide (SO2)	99%

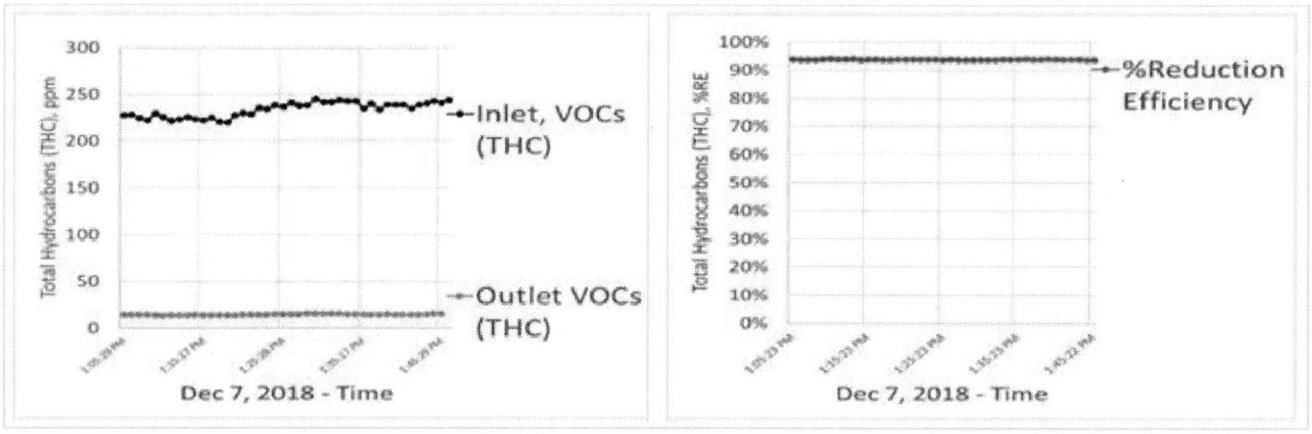

RESULTS OF VOC REDUCTION EFFICIENCY PERFORMANCE MEASURED BY FLAME IONIZATION DETECTOR (FBC DEMONSTRATION UNIT):
Achieved >90% Reduction Efficiency meeting PCWP MACT Standard[34]

---

[33] Emissions results FBC 2018 at commercial wood drying site done by Advanced Industrial Resources, Inc. Results were demonstrated during three 1-hour testing intervals. https://airtest1.com/ . (Accessed September 18, 2021)

[34] "Electronic Code of Federal Regulations (eCFR)". eCFR website. See https://www.ecfr.gov/cgi-bin/text-idx?node=sp40.13.63.dddd#ap40.14.63_12292.1 (Accessed April 16, 2020).

RESULTS OF VOLATILE ORGANIC COMPOUNDS CAPTURE:

Terpenes were collected from the FBC demonstration unit. These terpenes are a mixture of terpene hydrocarbons, primarily Alpha and Beta pinenes. These terpenes are responsibly-recovered, biobased materials. Potential applications for these terpenes include use in fuels, fragrances, solvents, adjuvants, cleaners, and polyterpene resins.

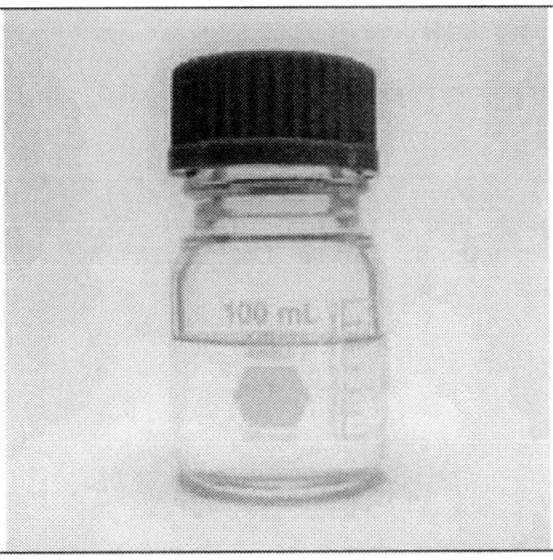

TERPENES COLLECTED FROM FBC DEMONSTRATION UNIT

VALUE OF TERPENES:

The Pine Chemicals Association has documented the gallons of terpenes per oven dry ton of wood in a variety of tree species. Typically, there is ~1 gallon of turpentine (terpenes)[35] per oven dry ton of wood product produced from green pine wood from the SE part of the United States, i.e. Loblolly Pine, Jack Pine, Slash Pine, Shortleaf Pine, Ponderosa Pine, Pitch Pine, Pond Pine, Red Pine, and Spruce Pine. If 75% of this is evaporated and captured from the drying process, the value of the terpenes per year can be calculated by multiplying this yield by the oven dry tons (ODT) of wood produced and a typical price per gallon[36] for terpenes. For example:

$$\left(\frac{1\ Gallon\ Terpenes}{ODT\ Wood}\right) * \left(\frac{ODT\ Wood}{Year}\right) * (\%Yield) * \left(\frac{\$\ Dollars}{Gallon\ Terpenes}\right) = \$\ /\ Year$$

TABLE VI
TERPENES TYPICAL PROPERTIES:

Property	Result
Appearance	Clear Liquid
Moisture	<0.1%
Weight per Gallon	7.2 pounds per gallon
Specific Gravity	0.86
Sulfur Content	< 10 ppm

---

35 "Sulfate Turpentine Recovery". Pulp Chemicals Association. Pages 9-12, 127. https://www.pinechemicals.org/store/viewproduct.aspx?id=1579188 (Accessed Jan 2, 2021).

36 "Price Chart for Gum Turpentine". PineChem website. http://www.pinechem.net/images/gall/ChartGNE.GIF Accessed Jan 3, 2021).

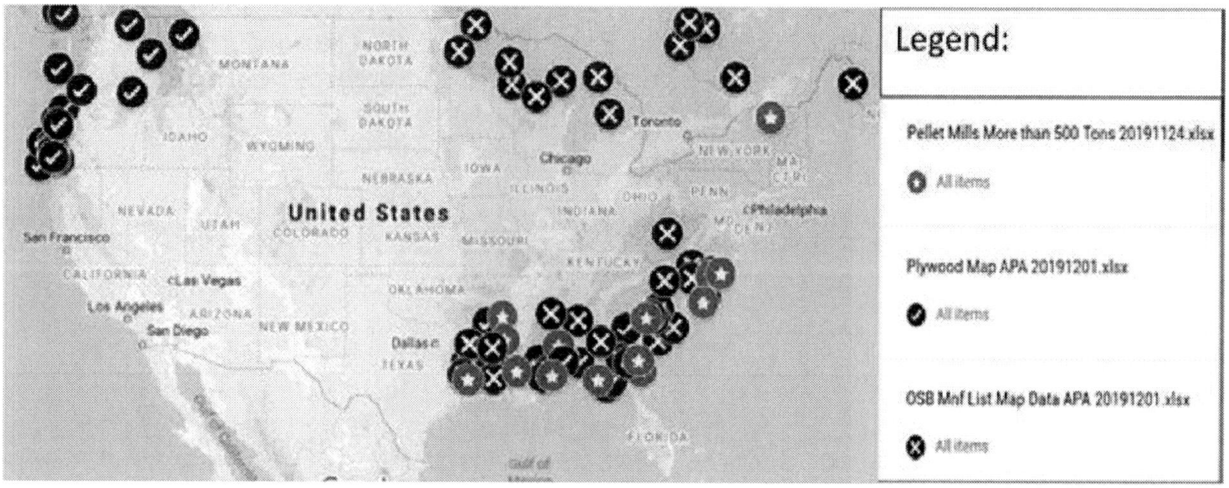

MAP OF POTENTIAL WOOD DRYING SITES FOR FBC USE:
This map shows potential sites where various wood products are produced where the FBC may be used to control emissions. Potential sites include but are not limited to Oriented Strand Board (OSB), lumber (sawmills)[37], plywood[38], wood to biofuels, i.e. wood pellets[39], particleboard, Medium Density Fiberboard (MDF), etc.

RESULTS OF DISCUSSIONS WITH ENVIRONMENTAL AGENCIES:

Official stack testing results were discussed with two state environmental agencies and two federal environmental protection agency (EPA) personnel. In all discussions, environmental agency personnel were supportive of providing a permit for full size commercial FBC unit.

FINAL RESULTS FROM WOOD DRYING DEMONSTRATION. THE UNIT:

- Provided salable terpenes, composed primarily of Alpha and Beta Pinenes.
- Reduced greenhouse gas (GHG) emissions.
- Reduced energy usage. It eliminated natural gas usage and reduced electrical usage.
- Met the PCWP MACT requirement of >90% Reduction Efficiency[40] in both tests. In the second test, the system met the compliance requirements three times in a row, over each of the three one-hour testing intervals.
- Demonstrated good operational reliability.
- Enabled State and Federal environmental personnel that reviewed these results to support permitting for a full commercial FBC system.

---

[37] "The mill map allows you to visually locate and drill down to find the mill you're interested in." Primary Forest Products Network website. https://primary.forestproductslocator.org/mill-map (Accessed September 18, 2021).
[38] "Manufacturer Directory". Engineered Wood Association website. https://www.apawood.org/manufacturer-directory (Accessed September 18, 2021).
[39] "Biomass Magazine". Biomass Magazine website. See http://biomassmagazine.com/plants/map/pellet/ (Accessed September 18, 2021).
[40] "Electronic Code of Federal Regulations (eCFR)". eCFR website. See https://www.ecfr.gov/cgi-bin/text-idx?node=sp40.13.63.dddd#ap40.14.63_12292.1 (Accessed April 16, 2020).

IX. CONCLUSIONS:

The FBC provides an affordable, reliable, and accessible clean technology for controlling VOCS, HAPS, and even particulate in some applications. This technology is well established in many applications including automotive and electronics manufacturing where it has been used successfully for many years. It is an emerging technology for use in biomass (i.e. wood drying) applications.

Especially for process exhaust streams with low VOC concentrations, the FBC technology is a patent[41] pending[42], cost competitive, industrial air pollution control technology that enables operators to advantageously:

1. Generate revenue via sales of recovered organic products (including biobased products such as terpenes in biomass applications) and sales of carbon offsets
2. Improve sustainability by reducing energy usage and thus greenhouse gas (GHG) emissions
3. Reduce operational costs by reducing energy usage (natural gas and electricity) and maintenance costs.
4. Enhance VOC mass transfer to the BAC media
5. Provide continuous in-situ BAC media regeneration and reactivation. This enables the media to continue to be reused.
6. Minimize water adsorbed on the BAC media thus maintaining its ability to adsorb VOCs, even in high moisture content process exhaust streams

Like any air pollution control technology installation, careful consideration of many design parameters, the unit's specific operating profile, and the operator's goals need to be taken into account during the design, installation, and startup/optimization processes. There are excellent resources, including Environmental C&C Inc.[43], Captis Aire LLC[44], and the Institute of Clean Air Companies (ICAC)[45] personnel, to assist with evaluating, implementing, and optimizing FBC applications.

---

[41] "METHODS AND SYSTEMS FOR RECOVERING TERPENE COMPOSITIONS FROM WOOD DRYING EXHAUST". World International Patent Office (WIPO) Website. https://patentscope.wipo.int/search/en/detail.jsf?docId=WO2019148208 (Accessed September 18, 2021).

[42] "METHODS AND SYSTEMS FOR CONTROLLING EMISSIONS FROM WOOD DRYING PROCESSES". WIPO Website. https://patentscope.wipo.int/search/en/detail.jsf?docId=WO2019148207 (Accessed September 18, 2021).

[43] "Environmental C&C Inc". Environmental C&C website. https://www.environmentalcc.com/ (Accessed September 18, 2021).

[44] "Emissions Control Technology Captures Emissions to Clean the Air". Captis Aire website. https://www.captisaire.com/ (Accessed November 21, 2021).

[45] "The Trusted Voice of the Clean Air Technology Industry". Institute of Clean Air Companies (ICAC) website. https://www.icac.com/ (Accessed September 18, 2021)

X. ACKNOWLEDGEMENTS:

The author recognizes inputs from a broad spectrum of personnel at the commercial OSB manufacturer and at Environmental C&C[46] including Harold Cowles, John Berger P.E., and Jim Starek.

CAPTIS AIRE CAPTURE-CONCENTRATE-CLEAN	Environmental C&C, Inc. Solvent Recovery & VOC Abatement Systems
Kim TUTIN Captis Aire LLC [47] 1612 W. Cleveland Ave, Suite 001 East Point, GA 30344-3213 Tel: +01 404 5802795, E-mail: kim@captisaire.com Website: www.captisaire.com	Harold COWLES Environmental C&C Inc. 16705 E. Calle de las Estrellas Road Scottsdale, AZ 85262-5982 Tel: +01 518 2250315, E-mail: halc@ecnc.com Website: https://www.environmentalcc.com/

[46] "Environmental C&C Inc". Environmental C&C website. https://www.environmentalcc.com/ (Accessed September 18, 2021).
[47] "Emissions Control Technology Captures Emissions to Clean the Air". Captis Aire website. https://www.captisaire.com/ (Accessed November 21, 2021).

631638 Conversion of Plastic Waste to Fuel: Assessment of Technologies and Economics

Abstract:

Plastics are extremely valuable to our society and are widely used across industries such as food, beverage, textile, healthcare, automotive, aerospace, construction, agriculture, and oil. The global production of plastics, accumulative from 1950 to 2017, is 8.3 billion metric tons, with 6.3 billion tons waste generated, of which only 9% has been recycled, 12% incinerated, with the remaining 79% landfilled or released into the environment. Mechanical recycling is used to recycle about 95% of plastic waste today, however, this method works well only for homogeneous and non-contaminated plastic waste, and for easily identifiable objects such as bottles and containers made of PET and PE. Chemical recycling can offer a solution in recycling many of the plastics which mechanical recycling fails to process. Chemical recycling involves the deconstruction of polymers to monomers for conversion back into polymers, or to intermediates that can be used as fuel or feedstock by the chemical industry. A number of chemical recycling technologies are being developed, and the processes are being tailored to handle specific feedstocks and/or produce desirable products. In this project we assess companies that are involved in the chemical recycling of plastic waste: their technology and/or key innovations, the feedstocks and product(s), and the current scale of operation. Further, we evaluate in this project the economics of chemical recycling with a specific interest on plastic-to-fuel conversion.

623136 Continuous Back to Monomer Recycling of PET from Composite Materials Using a Twin-Screw Extruder

Abstract:

The current Packaging Act requires a 63% quota for the mechanical recycling of plastics from 2022 in Germany [1]. In order to meet this requirement, new efficient recycling technologies are necessary. Many plastics, such as polyethylene terephthalate (PET), are used in combination with other polymers. Apart from energy recovery, no recycling process for such multilayer material is available and it is therefore largely lost to the resource cycle.

The research project revolPET® specifically addresses this issue through the development of a continuous process for recovering the monomers terephthalic acid (TA) and ethylene glycol (EG) in primary raw material quality from mixed PET-containing waste [2]. While literature mainly reports batchwise depolymerization of PET in stirred tank reactors [3, 4], the revolPET® process uses an extruder as a continuous chemical reactor.

This contribution presents experimental results on the successful depolymerization of PET in composite materials into a TA salt and EG. The high flexibility of the extruder offers a wide range of options for process design in terms of thermal, mechanical and chemical conditions, so that suitable process windows are identified depending on the feed material. Key to this technology, the twin-screw extruder allows the two reagents to be mixed as solids. This innovative process intensification enables high degrees of depolymerization at average residence times below 2 minutes. In the experiments, a PET depolymerization degree of $\geq$ 97 % was achieved. Systematic parameter studies reveal the influence of temperature, filling degree of the screw chamber and residence time on the successful PET depolymerization. Figure 1 presents typical results for degree of depolymerization and residence time at varied screw revolutions.

In addition to the technical aspects, the contribution provides insight into the ecological assessment of this new recycling process. In particular, the challenge of the exothermic reaction to identify an optimal process window is shown. This topic is examined in more detail against the background of scaling. The presented experimental results will be discussed in detail and placed in the context of the current political debate.

Literature

[1] § 16 Absatz 2 VerpackG

[2] www.revolpet.eu

[3] A.M. Al-Sabagh, F.Z. Yehia, Gh. Eshaq, A.M. Rabie, A.E. ElMetwally, *Egyptian Journal of Petroleum* (2016) 25, 53-64, doi: 10.1016/j.ejpe.2015.03.001

[4] A. Oku, L.-C. Hu, E. Yamada, *J. Appl. Polym. Sci.* (1997) *63*, 595-601.

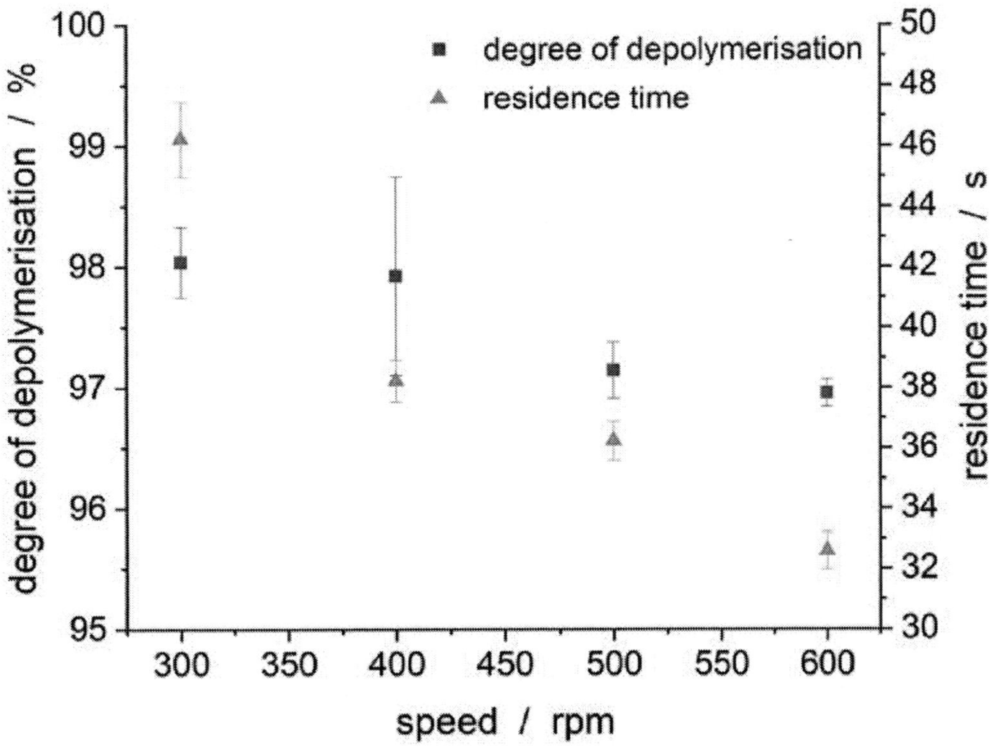

Figure 1: Degree of PET depolymerization in relation to the residence time at different speed

629642 The Effect of Residence Time and Heterogeneous Catalysis on Continuous Hydrothermal Liquefaction of Food Waste

Abstract:

The United States releases a significant amount of methane through the decomposition of waste in landfills; methane is a greenhouse gas with a global warming potential 25 times greater than carbon dioxide [1]. According to the EPA, landfilling accounts for approximately 17% of the total US methane emissions [1], and food waste is the largest non-recyclable component of municipal waste [2]. Although food waste currently harms the environment, it also presents an opportunity to solve another looming threat: the US dependency on non-renewable fossil fuels. Mainstream Engineering, in collaboration with Worcester Polytechnic Institute, is developing technology to convert food waste to energy through hydrothermal liquefaction (HTL) that simultaneously addresses our reliance on fossil fuels and the environmental impact of food waste.

HTL converts food waste, and other wet wastes, into an energy dense biocrude by reacting the food mixture of carbohydrates, lipids, and proteins in near supercritical water. Traditionally, HTL reactions have been uncatalyzed or used a homogeneous catalyst (i.e., Na_2CO_3). We have shown previously that using a solid oxide, heterogeneous catalyst, $CeZrO_x$, in a batch reactor improves biocrude yield and energy recovery [3]. We have subsequently studied the effects HTL residence time and catalyst addition in a continuous stirred tank reactor (CSTR). A model food waste slurry was fed to the CSTR via a dual high pressure syringe pump system. The pressure of the system was maintained using a back pressure regulator rated to 300 °C and 5,000 PSI. Commercially available $CeZrO_x$ powder was compressed into tablets and immobilized in the CSTR using a catalyst cage. The catalyst tablets were able to withstand the hydrothermal environment without physically disintegrating. The stability of the catalyst tablets was characterized using XRD and BET before and after HTL. The energy recovery, product yields, and biocrude properties (HHV, TAN, and CHNO) were determined to characterize the effectiveness of $CeZrO_x$ in a continuous HTL reactor and evaluate the impact of residence time on continuous HTL of food waste.

References

[1] EPA. *Greenhouse Gas Emissions*. [cited 2021; Available from: https://www.epa.gov/ghgemissions/overview-greenhouse-gases.

[2] USDA. *Why should we care about food waste?* [cited 2021; Available from: https://www.usda.gov/foodlossandwaste/why.

[3] Maag, A.R., et al., *Catalytic hydrothermal liquefaction of food waste using CeZrOx*. Energies, 2018. **11**(3): p. 564.

629415 Continuous Hydrotreatment of Sewage Sludge and *Spirulina* Algae Biocrudes from Hydrothermal Liquefaction: Different Catalysts for Different Organic Contaminants

Abstract:

Unprecedented global warming and raising public awareness are the driving factors to decarbonize long-haul transportation sector. In this regard, biomass could be a potential renewable source for the production of sustainable, carbon-neutral liquid fuels. Hydrothermal liquefaction (HTL) followed by catalytic hydrotreatment is one of the most promising pathways to covert varying feedstocks into drop-in fuels. The energetically dense black viscous product from HTL ("biocrude"), contains a diverse organic pool with considerable amount of oxygen (~5-18%), nitrogen (~1-8%) and metallic content (~0.1-0.6%). The presence of these organic contaminant results in poor miscibility with fossil fuels and poses new and exciting challenges e.g. thermal instability of biocrude at high temperatures which are indispensable for complete hydrodenitrogenation. This is why special attention is needed not only to realize this complex biocrude mixture but also to select different catalysts which performs best to saturate and remove certain organic contaminants under optimized conditions.

In this work, we utilized advanced analytical tools (i.e. FT-ICR MS etc.) and observed different classes of organometallics and nitrogen compounds from sewage sludge and *Spirulina* algae biocrude. From these results, we utilized three different types hydrotreating catalysts based on their porosity and nickel loading on γ-Al_2O_3 support and successfully demonstrate smooth continuous hydroprocessing operations (Figure 1) for hundreds of hours without coke formation, catalyst deactivation, and reactor plugging. Finally, 100% deoxygenation and ~96% denitrogenation were achieved. Furthermore, hydrotreated products underwent true-boiling point fractional distillation and showed on-spec metal content, density, pour point, cloud point and HHV for diesel and jet fuel.

.

Figure 1: Hydrotreated samples from the continuous hydrotreatment of sewage sludge (a) and *Spirulina* algae (b) biocrude.

Acknowledgement: This research has received funding from the European Union's Horizon 2020 research and innovation programme under grant agreement No 764734.

627420 Review and Modeling of Organic Removal in Anaerobic Sludge Blanket Bioreactors for Methane Production and Energy Recovery from Domestic Wastewater

Abstract:

The current domestic wastewater treatment paradigm centers on aerobic wastewater treatment technologies such as activated sludge. While effective in meeting U.S. EPA regulated discharge standards for organics and suspended solids, providing oxygen to aerobic microorganisms is energy-intensive, typically accounting for ~50% of wastewater treatment facility costs. In total, wastewater conveyance and treatment accounts for ~3% of our nation's electrical energy usage. Direct anaerobic treatment of dilute wastewater is a viable methane-generating alternative to aerobic wastewater treatment. Anaerobic microorganisms convert wastewater organics (fats, carbohydrates, proteins, etc.) to methane-rich biogas, which can be harnessed to create energy via technologies such as combined heat & power. If more energy is generated from the methane produced than is used to run the process itself, then the process can be considered "energy-positive" (or a net energy producer). Anaerobic sludge blanket bioreactors, most notably upflow anaerobic sludge blanket (UASB) bioreactors and anaerobic baffled reactors (ABRs), have been studied at bench- and pilot-scale for several decades. UASBs are also currently operated at full-scale in several locations worldwide, mostly in warmer climates. Despite the number of completed studies, no current review critically examines performance data from different sized sludge blanket bioreactors collected under varying conditions (e.g., varying wastewater temperatures and hydraulic residence times). Further, no current review study conducts predictive modeling (i.e., Monte Carlo Simulation) on aggregated sludge blanket bioreactor data to better understand the operation of such bioreactors under varying conditions. This study aggregates and examines data from 145 UASBs studies and 56 ABRs studies to understand bioreactor performance (e.g., chemical oxygen demand and total suspended solids removal) and methane generation. This study also employs Monte Carlo Simulation using @Risk to better predict bioreactor performance under varying conditions. Additionally, this study identifies critical gaps in literature (e.g., a lack of consistently reported methane generation) that may impact full-scale implementation in certain climates (e.g., cold weather climates). Last, results from this study can help inform decision makers to better understand of anaerobic sludge blanket bioreactors can meet regulatory standards or generate methane for renewable energy at their wastewater treatment facilities.

629993 Utilization of Marble Dust As a Concrete Additive to Enhance Its Properties

Abstract:

With the advancement in the building technology, each and every step to alter the cement properties for making it environmentally beneficial and equally strong is important. In this advancement, one always opts for an alternative which provides the strength to the building by reducing its cost and weight.

Marble often known as calcium carbonate ($CaCO_3$) has been a well known material used as a building material since the ancient times. Marble is often available in combination with Silicon Oxide (SiO_2), Aluminum Oxide (Al_2O_3), Iron Oxide (Fe_2O_3), Magnesium Oxide (MgO) and Calcium Oxide (CaO).

Marble is utilized at a number of operations and the waste obtained requires an extra effort for the adequate environmental disposal. The waste marble left in the environment can be harmful therefore must be utilized for an alternative process to reduce its harmful effect. The addition of marble dust to concrete is proving to be an effective alternative for the process. The process of marble addition is participating in the hydraulic reactions contributing significantly to the composition and microstructure of hydrated products. Another way of utilizing the marble dust is to utilize it in the geo-polymer concrete which will be strengthening the concrete properties.

The entire paper is evaluating the utilization of the marble dust in concrete so that the unsafe waste of marble is used and fewer efforts is to be taken for its disposal and also to accomplish viability, economy and environmental conservation.

629782 Deployment of Machine Learning Models in Pharmaceutical Development

Abstract:

The advantage of model driven decisions in research organizations has long been recognized and applied in the pharmaceutical research and development. From the conceptualization of the Quality by Design framework as a series of mathematical constructs [1,2] to the application of digital twins [3]. Machine Learning methodologies are widely applicable and efficient tools to construct mathematical models of relevant pharmaceutical processes enabling process optimization, robustness characterization, and failure rate estimation. However, the widespread adoption of these approaches requires a combination of streamlined availability, and institutional know-how to interpret and incorporate the results in the decision-making process.

This talk addresses different approaches to leverage emerging and stablished tools in Machine Learning to pharmaceutical development applications by comparing the adoption of specific algorithms and computational platforms. Specifically, a generic workflow for data exploration analysis, design of experiments, feature selection, and model exploration has been developed and deployed with a combination of cloud computing and Jupyter notebooks. In our experience this approach has been more successful at integrating a model-based decision making culture in process development than previously attempted alternatives.

[1] Peterson, J.J., 2008. A Bayesian approach to the ICH Q8 definition of design space. *Journal of biopharmaceutical statistics*, *18*(5), pp.959-975.

[2] Garcia-Munoz, S., Luciani, C.V., Vaidyaraman, S. and Seibert, K.D., 2015. Definition of design spaces using mechanistic models and geometric projections of probability maps. *Organic Process Research & Development*, 19(8), pp.1012-1023

[3] Chen, Y., Yang, O., Sampat, C., Bhalode, P., Ramachandran, R. and Ierapetritou, M., 2020. Digital Twins in Pharmaceutical and Biopharmaceutical Manufacturing: A Literature Review. *Processes*, *8*(9), p.1088.

626098 Unsupervised Learning from Sets of Data Using Contrastive Latent Variable Models

Abstract:

In unsupervised learning, dimensionality reduction is an important tool for data exploration and visualization. Because these aims are typically open-ended, it can be useful to frame the problem as looking for patterns that are enriched in one dataset relative to another. These pairs of datasets occur commonly, for instance a population of interest vs. control or signal vs. signal free recordings.However, there are few methods that work on sets of data as opposed to data points or sequences. Here, we present a probabilistic model for dimensionality reduction to discover signal that is enriched in the target dataset relative to the background dataset. The data in these sets do not need to be paired or grouped beyond set membership. By using a probabilistic model where some structure is shared amongst the two datasets and some is unique to the target dataset, we are able to recover interesting structure in the latent space of the target dataset. The method also has the advantages of a probabilistic model, namely that it allows for the incorporation of prior information, handles missing data, and can be generalized to different distributional assumptions. We describe several possible variations of the model and demonstrate the application of the technique to de-noising, feature selection, and subgroup discovery settings.

627704 Energy Dispersive X-Ray Hyperspectral Image Analysis and Chemometrics for Catalyst Characterization

Abstract:

Multi-dimensional images with chemical information embedded at pixel-level (so-called Hyperspectral Imaging) is the next generation of Image Analysis and Machine Vision. This advanced technology has attracted the attention of different industry sectors in the last two decades, ranging from metalurgic, mining and pharma, to high-value-added manufacturing industries such as biotechnology and biomedical imaging. Recently, the combination of Scanning Transmission Electron Microscopy (STEM) and Energy Dispersive X-ray spectroscopy (EDX) has been developed to characterize the surface area of catalysts in the chemical industry. The measurements are done by scanning the electron beam over the sample and at the same time measuring the X-ray spectrum – so-called spectral imaging [1]. The spectral dimension of the hyperspectral images consists of X-ray counts acquired at different energy channels. Chemical data is corrupted with noise, which comes from the time-dependent arrival of discrete particles on the sensor. This noise typically follows a Poisson distribution, which represents the probability of occurrence or events (X-ray counts) during a given period of time.

The application of Multivariate Image Analysis (MIA) techniques and Multivariate Curve Resolution (MCR) models becomes essential for the analysis of EDX hyperspectral images. This contribution proposes a modeling framework that permits segregating hyperspectral X-Ray images into simpler images (so-called Distribution Maps, DM's), which can be directly related to each of the chemical compounds present in the mixture [2]. From these DM's, chemical-textural score images (SI's) are further obtained. Finally, from all these DM's and SI's, different types of features, such as quantitative, morphological or textural can be extracted and combined into new data structures [3]. This new source of information is used to build multivariate statistical models for process understanding and prediction purposes at a MIA image-based level. This approach allows us to study similarities and differences between and within types of catalysts -i.e. different samples of the same catalyst across batches, and across locations, and the potential effects on quality properties of interest. To illustrate the chemometric framework for catalyst characterization, STEM-EDX images of real industrial catalyst will be used.

[1] R. Haswell, D. W. McComb and W. Smith, Preparation of site-specific cross-sections of heterogeneous catalysts prepared by focused ion beam milling, Journal of Microscopy (2003) Vol. 211, Pt 2, pp. 161-166
[2] J.M. Prats-Montalbán, A. de Juan, A. Ferrer, Multivariate image analysis: a review with application, Chemometrics and Intelligent Laboratory Systems, 107: 1-23, 2011.
[3] C Duchesne, JJ Liu, JF MacGregor, Multivariate image analysis in the process industries: A review, Chemometrics and Intelligent Laboratory Systems 117, 116-128, 2012.

634389 Applications of machine learning in production enhancement and field monitoring in Oil & Gas industry

Abstract:

Due to the fast advancement of machine learning technologies and availability of diverse types of data collection platforms, we have been seeing the growing machine learning applications in a wide spectrum of industries, spanning from emerging industries, e.g., autonomous driving, to traditional ones, such as Oil & Gas (O&G).

In this talk, we will discuss three different machine learning applications in the O&G industry. In specific, we will discuss a machine learning-based approach for understanding the multiphase flow system studied in the experimental fluid mechanics; which shows the potential of developing a complimentary methodology to assist the first principal physics; secondly, we will describe a deep learning-based object detection system that can be used to monitor the operation field or the natural environment, with the goal to improve operation safety and meet the regulation requirement ; Thirdly, we will introduce the content understanding of Process & Instrument Diagrams (P&IDs) using computer vision techniques.

Each of the above studies represents a unique usage scenario in the value chain of O&G industry. We would like to demonstrate the diverse types of machine learning applications and discuss their respective technical challenges that O&G industry presents.

AUTHOR INDEX

Abajorga, Hikma 115
Abdelhamid, Moaaz 79
Aboagye, Emmanuel A 131
Adams II, Thomas 129
Afarani, Hajar Taheri 4
Agbleze, Selorme 15
Ahmed, Isteaque 56
Ahuja, Sejal 241
Al-Atabi, Hayder 213
Al-Sharify, Zainab 213
Aldeghi, Matteo 81
Alexander, Ronald 157
Alexandridis, Paschalis 235
Alhameedi, Hayder 213
Alva, Carolina 79
Amar, Vinod 194
Anthony, Brian W. 136
Antonakoudis, Athanasios 153
Arbeiter, Florian 88
Armstrong, Cameron 176
Aspuru-Guzik, Alán 81
Attia, Lucas 5
Avadiappan, Venkatachalam 164
Azad, Mohammad 87
Babi, Deenesh K. 171, 173
Baca, Kalin 187, 190
Bademi, Apoorva 208
Badgwell, Thomas 72
Badr, Sara 51
Baker, Sarah 21, 22
Bandlamudi, Santosh Rathan Paul 191
Bangi, Mohammed Saad Faizan 63
Barros, Arianne 52
Bartling, Andrew 159
Bartusiak, R Donald 72
Barz, Kevin D. 119
Bayles, Alexandra V. 192
Beckingham, Bryan 118
Beckingham, Lauren 118
Bekaert, Bram 182
Benito, Marta Moreno 186
Bennett, Madelyn G. 190
Berger, John 215
Bernardi, Andrea 153
Bertran, Maria-Ona 171, 173
Bhakte, Abhijit 124
Bhethanabotla, Venkat 60
Biermann, Lars 236
Biernacki, Joseph J. 4, 77
Bispo, Heleno 15, 52
Bkour, Qusay 204
Black, Marcie 104
Blackwood, Daniel O. 186
Bohling, Peter 186
Bollas, George 66
Bonislawski, Arlan 111
Bortner, Michael 3, 75, 116
Boukouvala, Fani 165
Braatz, Richard D. 136, 172
Brenza, Timothy M. 119
Brepohl, Esther 236
Brettmann, Blair 76
Buehler, Markus J. 74
Burns, Mark A. 58
Buttazoni, Lucas 164
Cai, Lingrui 126
Cajner, Hrvoje 88
Camarda, Kyle 135
Cammarata, Paul 121
Campos, Gustavo 163
Carter, Michael J. 61
Casey, Eoin 180
Castello, Daniele 239
Cavagnaro, Robert 195
Celestine, Asha-Dee 118
Cetegen, Shaylin 205
Cha, Younghwan 204
Chea, John 131
Chen, Bingzhen 145
Chen, Yingjie 48
Cheng, Shuxing 245
Cherney, Samantha 111
Chiang, Leo 113
Cho, Young Tae 20
Cho, Youngtak 209
Choi, Jae-Soon 42
Choksi, Naitik A. 142
Chowdhury, Aniruddha 120
Chowdhury, Dhrubajit 68
Chundawat, Shishir 167
Cipriani, Ciera 24
Coley, Connor 152
Comparini, Edwin 126
Congdon, Taylor 235
Corzo, Carolina 79
Cowles, Harold 215

Crawford, Grant A. 61
Daniel, Fraser 1
Das, Arit 3, 75, 116
Davis Jr., James H. 191
Davis, Jonathan 21, 22
De Beer, Martin 58
De Beer, Thomas 182, 184
Deng, Daniel Z. 195
DeOtte, Joshua R. 21
Depaifve, Sebastien 78
Deshpande, Dnyanesh 160
Diao, Jinjin 40
Dinh, San 157
Diwekar, Urmila 128
Dombrowski, Katherine 165
Donovan, Katrina J. 119
Doshi, Pankaj 186
Dowling, Denis 180
Dudukovic, Nikola 21, 22
Duffie, Whytneigh 119
Dunn, Ryan 115
Duoss, Eric B. 21, 22
Dupuis, Dan 203
Durand, Helen 90, 95, 99, 111
Eder, Simone 88
Eichert, Carsten 236
El-Farra, Nael 97, 163
Elharati, Mohamed Ali 204
Ellis, Matthew 97
Esko, Iiro 109
Esmaeili, Mohsen 89
Etemadi, Martin 75
Fang, Nicholas X. 20
Farlessyost, William 208
Feng, Maoqi 42
Ferguson, Steven 180
Ferrer, Alberto 244
Filipova, Tsvetanka S. 119
Fisher, Julia 192
Fong, Erika 21, 22
Forte, Connor 6
Francqui, Filip 78
Fromen, Catherine 5
Gao, Haijing 245
Garboczi, Edward 4
Garcia-Munoz, Salvador 182
Geier, James 131
Gemeda, Hawi 21, 22
George, Kyle 89
George, Sherine 121
Ghijs, Michael 184
Glace, Matthew 176
Glaws, Andrew 159
Gleadall, Andy 1
Glezakou, Vassiliki-Alexandra 195
Goluch, Edgar D. 101
Gomes, Joseph S. 137
Grant, Stephen 106
Gray, McMahan L. 201
Grossman, Leon 109
Guan, Yuanfang 126
Gupta, Ram 193, 194
Gupta, Shashwat 48
Gyorgypal, Aron 167
Habibi, Sanaz 58
Haider, Muhammad Salman 239
Han, Gian 20
Han, Seulki 66
Hanselman, Christopher 126
Hasan, M M Faruque 188
Hassan, ASO 213
Haswell, R. 244
He, Q. Peter 122
Heldebrant, David 200
Helgeson, Matthew 192
Heller, Daniel 102
Helmbrecht, Hawley 138
Hendrix, Howard 165
Hernandez, Sergio 194
Herve, Philippe 19
Hickman, Riley J. 81
Hilborn, Haley 6
Hirl, Katharine 155
Hong, Moo Sun 172
Hou, Dianhao 60
Hwang, Sungwon 209
Häse, Florian 81
Ierapetritou, Marianthi 48, 167
Iftakher, Ashfaq 188
Izard, Anna Guell 21
Jajcevic, Dalibor 186
Jang, Gyoung Gug 23
Ji, Tuo 201
Jiang, Zhihua 122
Johnson, Brad 182
Jones, Christopher W. 165
Joshi, Prateek 114
Kaiser, Johann 171, 173
Karimi, Iftekhar 16, 148, 160
Katschnig, Matthias 88
Kazmer, David 115
Kedzierski, Brent 7
Kelsey, Katelyn 109
Kestering, Daniel 15

Khan, Zoheb 180
Khinast, Johannes G. 186
Kim, Semie 214
Kim, Seok 20
Kimbell, Georgia 87
Kimber, James 186
Kitchens, Christopher 127
Koker, Brian 115
Kong, Fanhe 165
Kotula, Anthony 57
Koutsamanis, Ioannis 79, 88
Kowsari, Kavin 20
Krarup, Janus 171, 173
Krühne, Ulrich 171, 173
Kshirsagar, Shivani 178
Kwon, Joseph 63, 107
Lakelin, Matthew 153
Lee, DongJin 209
Lee, Hojae 164
Lee, Kai 186
Li, Bowen 166
Li, Dan 146
Li, Guoqiang 120
Li, Jie 146
Lifeng, Zhang 145
Lim, Young-Il 214
Lima, Fernando V. 15, 52, 157
Linvill, Chelsea Q 240
Lischeske, James J. 159
Liu, Jian 195
Liu, Y. 202
Liu, Yu 163
Lively, Ryan 165
Lochmann, Dirk 79
Long, Xiaoyan 180
Lu, Amos E. 172
Lubongo, Cesar 235
Lumay, Geoffroy 78
Luo, Zhiyao 82
Mackley, Michael 131
Maheshwari, Dr. Utkarsh 241
Majumdar, Sudip 134
Mannan, Nasir 66
Mansell, Thomas J. 45
Manthiram, Karthish 196
Maravelias, Christos 164
Marnot, Alexandra 76
Martinez, Jose Maria Gonzalez 244
Matic, Marko 186
McClain, Monique 68
Meng, Xianzhi 43, 44
Mesbah, Ali 151
Messina, Dominc 99, 111
Meyer III, Harry M. 42
Mihalos, Aristidis 238
Misra, Shamik 164
Mohammad, Abdul Salam 77
Monjur, Mohammed Sadaf 188
Montalban, Jose Manuel Prats 244
Monteiro, Nina 129
Montgomery, Macrae 6
Monty-Bromer, Chelsea 103
Moon, Tae Seok 40
Morris, Ian 240
Moschetta, Eric 175
Moser, Newell H. 4
Mozhi, Anbu 86
MSc 79, 186
Mukherjee, Rajib 128
Murnen, Hannah 134
Muzzio, Fernando 48
Nachenberg, Andrew 109
Nagy, Zoltan 178
Nam, Sang-Hoon 20
Nance, Elizabeth 138
Narasimhan, Shilpa 97
Nash, Connor P. 203
Nasr-Esfahani, Ebrahim 4
Neveu, Aurelien 78
Nguyen, Manh-Thuong 195
Nieman, Kip 111
Nimlos, Claire 203
Nopens, Ingmar 184
Norinder, Jonas 14
Norton, M. Grant 204
Nowbahar, Arash 192
Ocampo, Ana Belén 79
Ojo, Gbemisola 135
Olawuni, Deborah 87
Olsen, Greta M. 190
Olsen, Greta 187
Orbey, Nese 6
Owens, James T. 3
Oyama, Henrique 90, 95, 99, 111
Ozbek, Sebnem 61
Padon, Mia 240
Palazoglu, Ahmet 163
Papadopoulos, Athanasios 92
Papadopoulos, Lazaros 92
Papageorgiou, Dimitri 55
Papathanasiou, Maria 153, 169
Park, Jay 115
Parrish, Christine 134
Paschetag, Mandy 236

Patel, Premal ... 5
Paulsen, Alex D. ... 238
Paulson, Joel ... 142
Peng, You ... 113
Pentzer, Emily ... 24
Peters, Andrew ... 120
Pfluger, Andrew ... 240
Pinelo, Manuel ... 171, 173
Pravin, P S ... 82
Pritchard, Zachary D. ... 58
Priye, Aashish ... 56
Prucka, William R. ... 126
Pu, Yunqiao ... 43
Qi, Jerry ... 6
Qin, S. Joe ... 54
Rabideau, Brooks ... 191
Radadia, Adarsh ... 1
Ragauskas, Arthur J. ... 43, 44
Rahman, Minhazur ... 111
Ramakrishnan, Srividya ... 178
Rangan, Keshav Kasturi ... 90, 95, 111
Rangarajan, Srinivas ... 166
Raudenbush, Katherine ... 167
Ravula, Prashanth ... 148
Realff, Matthew ... 165
Reardon, Kenneth ... 105
Reddy, Jayanth Venkatarama ... 167
Regan, John M. ... 155
Renner, Julie N. ... 199
Reyer, Sebastian ... 79
Reza, Toufiq ... 211
Ribeiro, Marcelo P. A. ... 157
Richard, Tom L. ... 155
Rittenberg, Leslie ... 65
Rizkin, Benjamin ... 175
Rodene, Dylan ... 193
Roman-Ospino, Andres ... 48
Roper, Thomas ... 176
Rosendahl, Lasse ... 239
Rothstein, Samuel ... 45
Ruckdashel, Rebecca ... 115
Ruddy, Daniel ... 203
Rushdi, Salih ... 213
Sachde, Darshan J ... 165
Sadati, Monirosadat ... 89
Sahinidis, Nikolaos ... 182
Salar-Behzadi, Sharareh ... 79
Sarkis, Miriam ... 169
Saunders, Steven ... 204
Schiffer, Zachary ... 196
Schmidt, Eric G. ... 121
Schneider, Hannes ... 236
Scholl, Stephan ... 236
Schreier, Marcel ... 197
Schultz, Guilhermina ... 157
Scott, Joseph K. ... 165
Scott, Michael ... 176
Scott, Timothy F. ... 58
Seferlis, Panos ... 92
Sen, Swastik ... 45
Severson, Kristen ... 243
Sexton, Andrew ... 165
Shadle, Lawrence J ... 15
Shafe, Anwar ... 120
Shah, Nilay ... 153, 169
Shahrukh, Mohd ... 148, 160
Shangguan, Ning ... 134
Shao, Yuyan ... 195
Shell, Katelyn ... 193, 194
Shende, Rajesh ... 194
Shi, Fan ... 201
Shiflett, Mark ... 187, 190
Shin, Dongil ... 18
Shin, Yongbeom ... 18
Shinde, Vinita ... 118
Shoukat, Daniyal ... 6
Shreve, Michael J. ... 155
Singh, Dr. Manpreet ... 241
Singh, Ravendra ... 46
Singh, Shweta ... 208
Singh, Suraj Prakash ... 16
Sipple, Patrick ... 242
Sitapure, Niranjan ... 107
Sitaraman, Hariswaran ... 159
Slama, Eyke ... 79
Smith, Alexander ... 71
Smith, Joseph D. ... 213
Smith, Kaylee ... 58
Song, Min-Kyu ... 204
Sorourifar, Farshud ... 142
Soudris, Dimitrios ... 92
Spitler, Rod ... 121
Spoerk, Martin ... 79
Spörk, Martin ... 88
Squires, Todd ... 192
Srinivasan, Rajagopalan ... 16, 124, 148, 160
Stanier, Charles ... 137
Stengel, Jake ... 131
Stickel, Jonathan ... 159
Stuber, Matthew ... 205
Sugiyama, Hirokazu ... 51
Sullivan, Katherine ... 56
Sultana, Al Ibtida ... 211
Sun, Weike ... 136, 172

Susarla, Naga Lakshmi Ramana 178
Suthar, Kerul 122
Tabora, Jose 242
Taheri-Qazvini, Nader 89
Tamblyn, Isaac 81
Tate, James 121
Tavernard, Antônio 52
Teoh, Jia Heng 86
Teran, Vincent 198
Thibodeaux, John 121
Thierry, David 165
Thompson, Joshua 23
Tian, Huijie 166
Timko, Michael T. 238
Tipton, Roger B. 60
Tiwari, Asit 126
To, Anh 203
Todd, Samuel 191
Tompsett, Geoffrey 238
Toson, Peter 186
Triantafyllou, Niki 153
Trogrlic, Martina 186
Tsouris, Costas 23
Tucker, David 15
Tutin, Kim 215
Tyrrell, Kathryn 111
Uddin, Md Mosleh 212
Valenciano, Lucia Matamoros 190
Valentine, Connor S. 192
Van Langenhove, Selien 184
Vandeputte, Tuur 184
Vandromme, Marie 184
Vanhoorne, Valérie 182
Vashishtha, Saumye 127
Vasilas, Nikos 92
Vasista, Bairi Sai 124
Venkatasubramanian, Venkat 141
Venugopal, Mamta 126
Verrier, Hugh 186
Vicente, John B. 72
Villez, Kris 68
Vogt, Bryan 117
Vohra, Shaan 193
Walker, Travis W. 61, 119
Wang, Chi-Hwa 86
Wang, Jin 122
Wang, Jun 137
Wang, Qiuming 201
Wang, Xiaonan 82
Wei, Peiran 24
Weller, Thomas M. 60
Wen, Zhiyou 212
West, Kevin 191
Wetzel, Eric D. 115
Wick, Collin D. 120
Wilfong, Walter C. 201
Willetts, Ian 7
Woodward, Ian 5
Wright, Mark Mba 212
Xiao, Xin 146
Xie, Xiaohong 195
Xu, Bo 84
Xu, Qiang 162
Yaari, Zvi 102
Yang, Bin 42
Yang, Martin 164
Yang, Ou 167
Yang, Xiaojun 42
Yenkie, Kirti 131
Yi, Shouliang 201
Yoon, En Sup 18
Young, Ethan 159
Yu, Li 162
Yuan, Zhihong 145
Zavala, Victor 71
Zhang, Bo 126
Zhang, Dongda 146
Zhang, Haitian 84
Zhang, Liping 146
Zhang, Nan 146
Zhang, Shuyang 43, 44
Zhao, Zhi-Min 43
Zhu, Qinqin 84
Zuercher, Philipp 51